AF883267

Green Energy and Technology

For further volumes:
http://www.springer.com/series/8059

Martin Nicholson

The Power Makers' Challenge

And the Need for Fission Energy

Springer

Martin Nicholson
Lot 3, Skyline Road
The Pocket
NSW 2483
Australia
e-mail: martin@energyinachangingclimate.info

ISSN 1865-3529 e-ISSN 1865-3537
ISBN 978-1-4471-2812-0 e-ISBN 978-1-4471-2813-7
DOI 10.1007/978-1-4471-2813-7
Springer London Heidelberg New York Dordrecht

Library of Congress Control Number: 2012934365

© Springer-Verlag London 2012
This work is subject to copyright. All rights are reserved by the Publisher, whether the whole or part of the material is concerned, specifically the rights of translation, reprinting, reuse of illustrations, recitation, broadcasting, reproduction on microfilms or in any other physical way, and transmission or information storage and retrieval, electronic adaptation, computer software, or by similar or dissimilar methodology now known or hereafter developed. Exempted from this legal reservation are brief excerpts in connection with reviews or scholarly analysis or material supplied specifically for the purpose of being entered and executed on a computer system, for exclusive use by the purchaser of the work. Duplication of this publication or parts thereof is permitted only under the provisions of the Copyright Law of the Publisher's location, in its current version, and permission for use must always be obtained from Springer. Permissions for use may be obtained through RightsLink at the Copyright Clearance Center. Violations are liable to prosecution under the respective Copyright Law.
The use of general descriptive names, registered names, trademarks, service marks, etc. in this publication does not imply, even in the absence of a specific statement, that such names are exempt from the relevant protective laws and regulations and therefore free for general use
While the advice and information in this book are believed to be true and accurate at the date of publication, neither the authors nor the editors nor the publisher can accept any legal responsibility for any errors or omissions that may be made. The publisher makes no warranty, express or implied, with respect to the material contained herein.

Printed on acid-free paper

Springer is part of Springer Science+Business Media (www.springer.com)

Foreword

The modern world is caught in an energy resource and climate change pincer. As the growing mega-economies of China and India strive to build the prosperity and quality of life enjoyed by citizens of the developed world, the global demand for cheap, convenient energy grows rapidly. If this demand is met by fossil fuels, we are headed for an energy supply and climate disaster.

The development of an eighteenth century technology that could turn the energy of coal into mechanical work—James Watt's steam engine—heralded the dawn of the Industrial Age. Our use of fossil fuels—coal, oil, and natural gas—has subsequently allowed our modern civilization to flourish. It is now increasingly apparent, however, that our almost total reliance on these forms of ancient stored sunlight to meet our energy needs has some severe drawbacks, and cannot continue much longer.

For one thing, fossil fuels are a limited resource. Most of the readily available oil, used for transportation, is concentrated in a few, geographically favored hotspots, such as the Middle East. Most credible analysts agree that we are close to, or have passed, the point of maximum oil extraction (often termed "peak oil"), thanks to a century of rising demand. We have tapped less of the available natural gas (methane), used mostly for heating and electricity production, but globally, it too has no more than a few more decades of significant production left before supplies really start to tighten and prices skyrocket, especially if we "dash for gas" as the oil wells run dry. Coal is more abundant than oil or gas, but even this has only a few centuries of economically extractable supplies.

Then there is climate change and air pollution. The mainstream scientific consensus is that emissions caused by the burning of fossil fuels, primarily carbon dioxide (CO_2), are the primary cause of recent global warming. We also know that coal soot causes chronic respiratory problems, sulfur causes acid rain, and heavy metals (like mercury) induce birth defects and damage ecological food chains. These environmental health issues compound the problem of dwindling fossil fuel reserves.

Clearly, we must unhitch ourselves from the fossil fuel-based energy bandwagon—and fast.

In the developed world (US, Europe, Japan, Australia, and so on), we have enjoyed a high standard of living, linked to a readily available supply of cheap energy, based mostly on fossil fuels. Indeed, it can be argued that this has encouraged energy profligacy, and we really could be more efficient in the mileage we get out of our transport fleet, the power usage of our fridges, lights and electrical appliances, and in the design of our buildings to reduce demands for heating and cooling. There is clearly room for improvement, and sensible energy efficiency measures should be actively pursued.

In the bigger, global picture, however, there is no realistic prospect that we can use less energy in the future. More than a third of all humanity, some 2.5 billion people, have no access to electricity whatsoever. For those who do, their long-term aspirations for energy growth, to achieve something equating that used today by the developed world, is a powerful motivation for development. As oil supplies dwindle, it must be substituted (or the technology changed) if we are to keep our vehicles going. Oil is both a convenient energy carrier, and an energy source (we "mine" it). In the future, we will have to create our new energy carriers, be they chemical batteries or oil-substitutes like methanol or hydrogen. On a grand scale, that is going to take a lot of extra electrical energy!

Then there is the growing human population (which we hope will stabilize by mid-century at less than 10 billion) and the burgeoning impacts of climate change and other forms of environmental damage, there will be escalating future demands for clean water (at least in part supplied artificially, through desalination and wastewater treatment), more intensive agriculture which is not based on ongoing displacement of natural landscapes like rainforests, and perhaps, direct geo-engineering to cool the planet, which might be needed if global warming proceeds at the upper end of current forecasts.

In short, the energy problem is going to get larger, not smaller, at least for the foreseeable future.

Let us say we aim to have largely replaced fossil fuels with low-carbon substitutes by around the middle of this century. How can we meet this enormous challenge? Nuclear power based on atomic fission is one possibility. Renewable energy sources which harness solar energy and its derivatives are another. Improved efficiency in the way we use energy may also offer a partial fix, at least in the short term. In the broader context, however, to imagine that the global human enterprise will somehow manage to get by with less and accept an unreliable supply, just does not stack up with the reality of a fast-developing, energy-starved world. Put simply, citizens in Western democracies are simply not going to vote for governments dedicated to lower growth and some concomitant critique of consumerism, and nor is an authoritarian regime such as in China going to risk social unrest, probably of a profound order, by any embrace of a low growth economic strategy. As such, reality is demanding, and we must carefully scrutinize the case for *all* of the alternative energy technologies available to us.

Martin Nicholson, in *The Power Makers*, does this admirably—in a lucid and engaging way.

Technically, economically, socially, and politically—we face many challenges in trying to harness non-fossil fuel energy on a large scale. This is a complex topic—indeed; it requires a book like this to explain properly! How do we match supply to demand? What are the implications of harnessing diffuse energy sources? How do we manage energy flows that are variable and intermittent—sometimes delivering a lot of power, sometimes a little, and at other times none at all? What are the methods we can use to store large amounts of energy, to cover the non-generating periods? Is nuclear fission a clean and sustainable energy source? What are the costs and timescales required for a major energy transition?

Read on, and you will find the answers to these questions, and much more besides. It is a wonderful book.

Adelaide, 2 Dec 2010

Prof. Barry W. Brook
School of Earth and Environmental Sciences and Centre for Energy Technology The Environment Institute, The University of Adelaide

Preface

Modern society needs reliable electricity at an affordable cost.

I am in my sixties now and I have enjoyed reliable electricity for most of my life. Sure, there was the occasional brown-out or black-out, particularly in my youth. They were kind of fun. We could light candles or sit by the light of the open fire when the power went off. They were no fun for commerce and industry that had to close their doors. No fun for their workers who could not work and were sent home.

If you live in a western city today, you hardly ever see an interruption to your power supply. I would like to see it stay that way.

As a child, I never wondered where the power came from that worked the lights, the refrigerator, and the radio. When you flicked the switch it was just always there.

I am sure there are many people today who still do not know where their electricity comes from. They have not needed to know. It is just there!

Well, I hate to worry you, but that may be about to change.

All my life, nearly all my electricity has come from burning coal. That would be true for many of us all over the world. It is the cheapest source of electricity. Problem is, burning this coal produces air pollution and greenhouse gases (GHG) like carbon dioxide which is causing changes to the climate. Stop burning this coal and you stop producing the carbon dioxide but the lights go out unless you make the electricity some other way.

How easy is it to change the way we make electricity? Will we still have power to our homes and workplaces whenever we want it? Will we still be able to afford to pay the electricity bill? Well maybe, but maybe not—depending on how we finish up making the electricity. These are the questions we will explore in this book.

Low-carbon and affordable electricity are going to be key planks of any politician's agenda over the next few years. Renewable energy sources are widely seen as a key part of replacing dirty coal. We already have a pledge from the US President for 10% of their electricity to come from renewable sources by 2012, and

25% by 2025. The European Union (EU) has a target of 20% by 2020 as do China and my home country, Australia.

Decarbonizing electricity is going to happen, eventually, but which solutions make sense today and which do not?

One of the most significant barriers to replacing coal with most renewable energy sources is that renewable energy is relatively dilute and variable in supply. It took immense pressure over millions of years to create those energy-rich fossil fuels like coal from plant matter. Extracting the equivalent energy from "new" energy sources is proving to be difficult and expensive.

Renewable energy is becoming a bit like a religion with its own high priests who usually head up renewable energy companies or conservation organizations. Renewable energy is seen by some as *the* solution to replacing dirty coal and delivering low-carbon electricity. The savior from the ravages of climate change. Along with religion often comes zealotry claims supported by misinformation.

Many still see the problem as more political than technical. "We just need to fix government policy decisions", they say, blaming the coal, oil, and nuclear power lobby groups while ignoring the significant limitations of current generation renewable energy solutions. Not the least of which is cost.

Both sides of the debate (traditional sources vs. renewables) are at fault—each by over-stating their own case. The aim of this book is to inject some realism into the debate about making our electricity low-carbon while keeping the lights on and the electricity bills manageable.

Looking to a future where computers and other electric appliances dominate nearly every aspect of economic life, electricity is likely to be even more essential. Even with transport, where oil has enjoyed a near monopoly for over a century, electricity is making tentative inroads with electric cars and improved batteries. The future is electric.

We cannot indefinitely keep using up the Earth's stored energy resources like coal and oil faster than the Earth can replace them. Eventually they will run out and there will be no more concentrated stored energy left. Sustainable energy is a must. The real question is: how quickly can we transition from the stored energy sources of coal, oil, and gas to truly sustainable energy resources? And what resources are sustainable anyway?

Fission is the richest source of energy readily available on the Earth today. Natural radioactive decay takes place deep in the earth producing hot aquifers that we use to produce electricity and heat buildings in some places. But this natural radioactive decay only supplies a small fraction of the energy we need.

For over 50 years, man-made fission energy has made a significant contribution to meeting our electricity needs. It has turned out to be the safest and cleanest way to produce vast quantities of cheap electricity. Much safer than coal, gas, oil, or renewable hydro and even safer than many of the other renewable sources. Much cleaner than coal, gas, and oil, producing no carbon dioxide in operation. Fission can also be made to be sustainable, unlike fossil fuel stored energy sources. So why do we not just use more man-made fission energy? Unfortunately, there are influential lobby groups who do not see it this way.

Let us come clean right up front. Fission energy is just another name for nuclear energy. So why rename it? For many, nuclear power has bad vibes. Some will go to any lengths to stop its use. Nuclear power reminds them of nuclear bombs and nuclear wars. Nuclear bombs mean millions of innocent people dying from radiation sickness. Some think that because nuclear energy involves radiation so nuclear energy must be dangerous too. This is not so. The time has come to decouple this (often unconscious) association; hence the name change. Call it marketing spin if you like.

Man-made fission energy is the best option to address climate change. Not only is it the best option—it is the *only* option if we are to avoid expensive changes that disrupt our electricity supply. Fission is clean, green, and affordable.

All bold statements you might think, so they need to be defended. For those that doubt this can all be true, it is fine to be sceptical. Why not check these statements again after you have finished reading this book? If I have done my job properly, you should be nodding your head next time around.

I have titled this book *The Power Makers' Challenge* so I had better tell you who the Power Makers are. These are the people we trust to build, run, and control our electricity system and ensure the electricity gets safely and reliably to our homes and workplaces—the generator operators, the system operators and the network operators. For many years in most countries the Power Makers all worked for governments. Now many work for private businesses but the governments maintain a regulatory role—as they should; electricity is an essential service that we mess with at our peril.

The challenge facing our Power Makers is reliable, clean electricity. With over 65% of our electricity currently coming from fossil fuels (coal, gas, and oil) that produce 44% of the world's greenhouse gas emissions, this is no small challenge.

The Power Makers have no choice but to either clean up the fossil fuel power plants or replace them. As we will see in the book, they will probably do a bit of both. Neither is straightforward. Both are likely to be more expensive. And some of the replacement solutions are much better than others.

The Power Makers face this challenge at a time of ever increasing demand for electricity. The continuously rising world population; the moves by third and second world countries to become first world countries; the unsustainable reliance on oil for transport all guarantee a greater need for electricity everywhere for the foreseeable future.

This means alternative fuels need to reliably replace the fossil fuels and be sufficiently scalable to meet this future demand.

As we work through this challenge, we will see that many of the popular alternatives proposed are either not up to the task or will be so expensive when deployed on the scale needed that they will prove to be unattractive replacements. Hopefully, we can dismiss some of the popular myths about alternative energy sources for generating electricity as well as some of the popular myths about fission energy.

I have structured the book into five parts:

Part I Explains what electricity is and how we make it.

Part II Looks at the pluses and minuses of renewable energy technologies.

Part III Does the same for clean coal technologies.

Part IV Covers fission energy, how it works and answers the tricky questions.

Part V Looks at the likely future of electricity production.

There are a number of technical appendices and literature references for further reading for those who want a more in-depth understanding of the technologies. It is not necessary to read these to understand the book but they provide some useful reference material for those that have an interest. They might also answer many of those nagging questions you may have about the material.

Martin Nicholson

Acknowledgments

After writing my last book, *Energy in a Changing Climate*, it became clear to me that the energy carrier we call "electricity" was critical to establishing and maintaining our standard of living. Electrification of our homes and workplaces has seen changes to our world that would have been inconceivable in the nineteenth century, but generating electricity has also been a major contributor to man-made greenhouse gas emissions.

Two special fellow Australians have offered invaluable support to me while writing this book about the challenges faced by our electricity providers. Dr. Tom Biegler, who has many years' experience with energy systems and Professor Barry Brook who is Director of Climate Science at The Environment Institute, University of Adelaide have both extensively edited the book. Barry also has a strong interest in energy systems, particularly as they relate to the environment. I thank them both for their invaluable contribution to this book.

I also thank Dr. Gene Preston who works in the power industry in Texas for his first-hand experience with electricity generation and distribution in the US. Gene has provided many valuable suggestions for this book.

Acknowledgments

[illegible]

Contents

Abbreviations

ABWR	Advanced boiling water reactor
AC	Alternating current
AEMO	Australian energy market operator
BEV	Battery electric vehicle
BWR	Boiling water reactor
CAES	Compressed air energy storage
CCGT	Combined cycle gas turbine
CC	Combined cycle
CCS	Carbon capture and storage
CFL	Compact fluorescent lamp
CHP	Combined heat and power
CO_2	Carbon dioxide
CO_2-e	Carbon dioxide equivalent
CSP	Concentrated solar power
DC	Direct current
DG	Distributed or decentralized generation
DOE	US Department of Energy
EGS	Enhanced or engineered geothermal system
EIA	Energy information administration
EOR	Enhanced oil and gas recovery
EPBT	Energy payback time
EPR	European pressurized reactor
EU	European Union
EWEA	European Wind Energy Association
FBR	Fast breeder reactor
GDP	Gross domestic product
GHG	Greenhouse gas
GW	Gigawatt (power)
GWh	Gigawatt-hour (energy)
ha	Hectare
HEV	Hybrid electric vehicle

HVDC	High voltage direct current
IAEA	International Atomic Energy Agency
ICE	Internal combustion engine
IEA	International Energy Agency
IFR	Integral fast reactor
IGCC	Integrated gasification combined cycle
IPCC	Intergovernmental Panel on Climate Change
ISA	Integrated sustainability analysis, University of Sydney
ISCC	Integrated solar combined cycle turbine
ITER	International thermonuclear experimental reactor
km	Kilometer
kV	Kilovolt
kW	Kilowatt (power)
kWh	Kilowatt hour (energy)
LCOE	Levelized cost of electricity
LOLE	Loss-of-load event
LOLP	Loss-of-load probability
LRMC	Long-run marginal cost
LWR	Light water reactor
MIT	Massachusetts Institute of Technology
MJ	Megajoule
MSW	Municipal solid waste
MW	Megawatt
MWe	Megawatt electrical (power)
MWh	Megawatt hour (energy)
NEEDS	New Energy Externalities Developments for Sustainability
NG	Natural gas
NREL	National Renewable Energy Laboratory (US)
OC	Open cycle
OECD	Organisation for Economic Co-operation and Development
OCGT	Open cycle gas turbine
PBMR	Pebble bed modular reactor
PEV	Plug-in electric vehicle
PHEV	Plug-in hybrid electric vehicle
PHS	Pumped hydro storage
PHWR	Pressurized heavy water reactor
Pu-239	A fissile isotope of plutonium
PV	Photovoltaic
PWR	Pressurized water reactor
RE	Renewable energy
SCADA	Supervisory control and data acquisition
SFR	Sodium-cooled fast reactor
SMR	Small modular reactor
SRMC	Short-run marginal cost
STE	Solar thermal electric

Th-232	A fertile isotope of thorium
U-235	A fissile isotope of uranium
U-238	A non-fissile (fertile) isotope of uranium
UK	United Kingdom
US	United States
W	Watt (power)
WEC	World Energy Council
Wh	Watt-hour (energy)

Part I
About Electricity

The top-rated engineering improvement to the life of earthlings in this century was electrification. The majority of the top 20 achievements [of the twentieth century] would not have been possible without electricity. Electrification changed the country's economic development and gave rural populations the same opportunities and amenities as people in the cities. It provides the power for small appliances in the home, for computers in control rooms that route power and telecommunications, and for the machinery that produces capital goods and consumer products. If anything shines as an example of how engineering has changed the world during the twentieth century, it is clearly the power that we use in our homes and businesses.

Armstrong N (2000) The Engineered Century. National Press Club.

Neil Alden Armstrong was, of course, the first man to set foot on the moon. A task that would have been impossible without electricity.

Before we spend too much time thinking about how we could make electricity some other way that would not warm the planet, it is pretty important that we understand what electricity is and how we make it now. This is what we cover in this first part of the book.

Chapter 1
What is Electricity?

This chapter is for those with little knowledge about electricity. We will discuss electricity in the simplest terms and show how it is measured. We will then compare it to that other common commodity used in most households—water. For those who think electricity flows out of a switch on the wall in the way water flows out of the tap on the sink, then this chapter explains the difference. We also discuss the issue of electricity storage and the difficulties that present for the Power Makers. Finally, we discuss the Power Makers' challenge in providing us with a reliable electricity supply.

1.1 Electricity: Simply Expressed

"Electricity", in the way we are going to refer to it, is the flow of electrical power or charge along a conductor—typically a copper or aluminum wire. Electrons in the conductor can be made to pass an electric charge along the wire. This flowing electric charge is what we call an "electric current" or just electricity. Electricity needs a closed loop or circuit to flow as shown below.

Figure 1.1 shows a simple circuit that we would find in a battery torch (or a school physics lab). The lines shown in this diagram are wires or connectors that attach the battery to the switch and the lamp.

In this torch, the electricity is coming from the battery which is keen to move electrons along the wire as soon as we complete the circuit by closing the switch. When we turn on the torch switch, electrons start to flow through the wire and through the light bulb in the direction shown by the arrow.

For an incandescent lamp (the ones with a glowing element) like the one in the illustration, the electrons flowing through the filament in the lamp cause it to glow brightly, giving us light. The filament tries to block the electrons but they push their way through causing it to heat up and glow. Turning off the switch breaks the circuit and the electrons stop flowing so the filament stops glowing.

M. Nicholson, *The Power Makers' Challenge*, Green Energy and Technology,
DOI: 10.1007/978-1-4471-2813-7_1, © Springer-Verlag London 2012

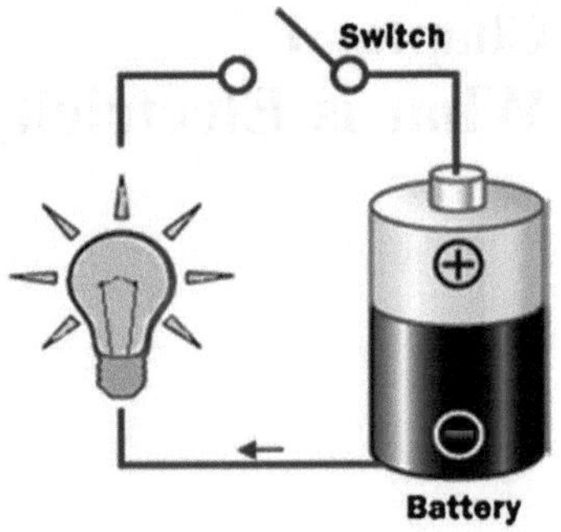

Fig. 1.1 Battery torch circuit

Of course for most of us the electricity in our homes does not come from batteries but generators. These generators might be many hundreds of kilometers away or in our own back shed. We will discuss generators in Chap. 2. For the electrical energy to get from the generator to our home it needs to travel along a wire and we will discuss electricity transmission in Chap. 3.

1.2 How Do We Measure Electricity?

There are a number of measurement units used with electricity including volts (V), amps (A), watts (W), and watt-hours (Wh). For the most part, we will restrict ourselves to watts and watt-hours so these are the units we will look at now. But first we need to differentiate between power and energy. Energy is the *amount* of electricity used (or produced) whereas power is the *rate* of consumption or production of that energy. Let us try to explain that a bit more clearly.

A popular (although incomplete) analogy for energy and power is to first talk about water volume and water-flow. We are all familiar with water volume. A liter bottle of water has a volume of 1 L—it is the *amount* of water in the bottle. Water-flow is the *rate* the water comes out of the bottle. If we put a hole in the cap of the bottle and turn the bottle upside down it might take 1 min for all the water to run out. This would be a flow rate of 1 L/min. For those mathematically inclined:

$$\text{Flow} = \frac{\text{Volume}}{\text{Time}}$$

"Energy" is like water volume (or the amount of water) and "power" is like water-flow. Substituting "power" for "flow" and "energy" for "volume" in the above equation then:

$$\text{Power} = \frac{\text{Energy}}{\text{Time}}$$

or more commonly: energy $=$ power $\times$ time.

Perhaps maths was not your strong suit at school so here is a concrete example that can help. A typical home toaster consumes electrical power at a *rate* of 1,000 W—called a kilowatt. If the toaster is left on for 1 h it will consume 1 kWh of electrical energy (as well as burn the toast). We have now introduced the key units of measure for electricity. For those who really like numbers, Box 1.1 will give you more information about W and Wh.

Box 1.1 Joules, Watts, and Watt-Hours

The standard international unit of energy is called a "Joule." The watt is a unit of power defined as a joule expended in one second. Therefore, a kilowatt or 1,000 W is the same as 1,000 J/s. There are 3,600 s in an hour so a kilowatt-hour is 3,600 × 1,000 J or 3.6 million joules. As you can see the joule is a very small unit of electricity so it is rarely used. Even a watt-hour is a relatively small unit of energy so we will stick with kilowatt-hour (kWh) as our preferred unit of electrical energy. Let us look at some more examples.

A typical incandescent electric light uses 40–100 W of power. An electric swimming pool pump uses 1,000 W—called a kW. The amount of energy we use depends on how long we use the light or the pump. Running a 1 kW pump for 1 h uses 1 kWh of energy, so a kWh is a unit of energy and a kW is a unit of power.

A watt is also a relatively small unit, so we talk of kW for 1,000 W, megawatt (MW) for 1 million watts (or 1,000 kW) and gigawatt (GW) for 1 billion watts (or 1,000 MW). I am sorry these are all a mouthful but these units will crop up throughout the book so it is important that they are introduced early on. To help you, there is a reference table in Appendix A showing how these units relate to each other.

1.3 The Big Differences Between Electricity and Water

Water and electricity are commodities that most of us in developed countries expect to be readily available in our home. They are both "on tap"–always there when we need them. They are both "piped" to our home from somewhere—even if we have no idea where that "somewhere" is.

For many people that is all they need to know. If the electricity or water stops flowing for some reason, they just ring the utility that supplies the water or the electricity and, hopefully, someone comes along and fixes the problem.

These obvious similarities might seduce us into thinking that water and electricity are very similar, but unfortunately that is not the case. Water occurs naturally in vast quantities. It falls from the sky when it rains or snows and flows into our reservoirs and dams. From there it travels through pipes, often by gravity, to our homes or workplaces, sometimes stopping along the way at some intermediate storage place like a local water tank.

Electricity also occurs naturally (in lightning for example) but in relatively small quantities and not in a way we can readily use it. All the electricity we use in our homes and workplaces is usually manufactured some distance away but occasionally it is produced locally. It needs to be manufactured using what is called a "primary" energy source. The most commonly used primary energy source is coal but we also use oil, gas, wood, water, air, sunlight, and uranium. Electricity is actually a "carrier" of energy from the primary source to its final place of use which might be your refrigerator.

Water can be readily stored in huge quantities and relatively cheaply. The rain does not always fall (thank goodness) so we build sufficient storage in dams, reservoirs and tanks to ensure a steady, continuous flow of water to our properties between expected rain periods. Sometimes in droughts, the water storage gets very low and we have to conserve water to make sure we do not fully drain the storage. As long as sufficient rain falls in a catchment area to serve the community over time we do not really care about when it falls as we can always build more storage to last through longer droughts if need be. Electricity, on the other hand, is very expensive to store—even in small quantities. We cannot just let it flow into a tank like we can with water.

Electricity (or more precisely electrical energy in the form of reactive chemicals) can be stored in batteries, even enough to run a home for several hours, but at significant cost and maintenance. Electrical energy can effectively be stored in larger quantities by using water storage. Electricity can drive pumps to push water from a large reservoir in a valley up to a reservoir on a hill. The water can then be allowed to flow back down again and converted back into electricity on the way down using spinning turbines that are a modern version of water wheels. But this is also very expensive, in terms of both construction costs and energy losses, and only a small fraction of the total electricity we produce is ever stored this way.

With water we really do not care when it actually fell to the ground because it is always stored somewhere before we use it. With electricity, every watt-hour we use has to be made at *exactly the same time* as we use it. With water, if more water falls than we can use, then we just store more of it or let it flow out to the sea.

1.4 The Power Makers' Challenge

Because we have very little electricity storage, electricity supply has to exactly match electricity demand, moment by moment. This is the major challenge for our Power Makers. This supply/demand challenge can be summarized as follows:

- The level of likely demand for electricity needs to be predicted with reasonable accuracy.
- Sufficient electricity supply needs to be available to meet the anticipated highest demand (known as the "peak demand").
- The amount of electrical energy generated at any moment needs to exactly match the demand at that moment.
- The "poles and wires" that bring the electricity to where we use it need to be the right size to handle the peak demand expected of them in their location.
- All this has to be achieved at the lowest possible cost.

These challenges will crop up and be discussed throughout this book but first let us talk about how the Power Makers make electricity.

Chapter 2
Faraday's Miracle

In this chapter we discuss how electricity generators work. Michael Faraday was the father of electricity generation. Almost 200 years ago, he discovered electromagnetic induction. Induction is the principle behind our modern electricity generators and transformers that are the backbone of our electricity networks. We will discuss, in simple terms, how generators work. We then discuss the different types of engines that can drive generators and the different energy sources used by the different engines. Finally, we compare these different engine types and introduce the term "capacity factor" which is critical to understanding the difference between the engines.

2.1 Faraday's Contribution

Faraday was born to a relatively poor family in London in 1791. He was a largely self-educated chemist and physicist who loved experimenting. His breakthrough experiment involved wrapping two coils of long insulated wire around an iron ring, and discovering that when he passed an electric current through one coil, another current was 'induced' in the other coil. Hence the term 'induction'. This experiment demonstrated the principle of electromagnetic induction and actually produced the first transformer.

He then took the idea further and discovered that a current could also be induced in the coil if it was moved over a stationary magnet. Faraday built the first electric dynamo by rotating a coil of insulated wire in a strong stationary magnetic field, thus converting rotating mechanical energy into electrical energy. This dynamo was the ancestor of modern power generators. We, and the Power Makers, owe Faraday a deep debt of gratitude.

M. Nicholson, *The Power Makers' Challenge*, Green Energy and Technology,
DOI: 10.1007/978-1-4471-2813-7_2, © Springer-Verlag London 2012

2.2 Generators Explained

As Faraday discovered, when a coil of insulated copper wire is rotated between two stationary magnets an electric current is induced in the coil. Many of us might have had (or still have) a dynamo on our bicycle wheel that operates the lights. This is how those small dynamos work.

A generator at a power station uses the same principle but generally uses an electromagnet to produce the magnetic field rather than a traditional fixed magnet. An electromagnet is just a magnet produced using electricity in a coil. The rotating part of a generator is called the 'rotor' and the stationary part is called the 'stator'. The electromagnet can be in the stator or the rotor. For a more technical description of how a synchronous generator works see Appendix B.1.

The electric current produced by power generators is normally an alternating current (AC), which means the current constantly changes direction, usually 100 or 120 times a second (50 or 60 cycles per second) depending on the country we live in. This is the form of electricity we receive from the power outlets in our homes and workplaces. The big advantage of AC is that the voltage can be readily changed using a transformer.

2.3 What Drives the Generator?

The generator shaft needs to be rotated by an engine of some kind. Commonly used engines are steam turbines, gas turbines (like jet engines on planes), internal combustion engines (like we use in our cars and trucks), wind turbines or water turbines. These engines convert energy from a fuel into rotary motion and the generator converts that rotary motion into electrical energy.

Most of the world's electricity is generated using steam turbines as the engine—see Fig. 2.1. Steam turbines need a lot of high pressure steam. The high pressure steam can be produced in a steam generator using some kind of heat source to heat the water to a very high temperature and pressure. Steam turbines take several hours to start-up from cold so they cannot be started quickly if needed.

A gas turbine works in a similar way to a steam turbine except it uses pressurised gas to spin the turbine blades rather than high pressure steam. A jet aircraft engine is a form of gas turbine. Most gas turbines produce their own pressurised gas by burning a fuel like propane, natural gas (NG) or jet fuel. The heat that comes from burning the fuel expands air and the high-speed of this hot air spins the turbine. Gas turbines can be started from cold in 10–20 min. Sometimes steam turbines are used in combination with gas turbines by using the waste heat from the gas turbine to generate steam for the steam turbine. These are called 'combined cycle gas turbines' or CCGT for short.

Water turbines used in hydroelectric dams have rotating blades like steam turbines but use fast running water to push the blades round. The blades in a water turbine look quite different to the blades in a steam turbine as water is much denser

Fig. 2.1 750 MW generator with 2 steam turbines. From Campbell G (1993)

than steam and moves much more slowly, but the principle is the same. Like gas turbines, water turbines can be started in a few minutes.

Wind turbines are rotated by the wind. These turbines look nothing like steam, gas, or water turbines. As the wind moves relatively slowly, wind turbines need huge blades—see the photograph Fig. C.1 in Appendix C. Wind turbines work intermittently depending on when the wind is blowing.

The engine we are all most familiar with is the internal combustion petrol or diesel engine we use in our cars and trucks. The same engines can be used to drive a generator. A portable petrol generator is made up of a small petrol engine attached to a generator that can produce mains electricity. Some grid electricity is also produced using large diesel engines driving generators.

Whatever the engine that drives the generators, be it steam, gas, water, wind or diesel the structure of the generator is very similar in all cases. What can be different is the speed of rotation and type of generator used depending on the engine used. See Appendix B.1 and Appendix C.5 for more detailed discussion on generator types.

2.4 What Powers the Engines?

All engines need a primary energy source. These energy sources are even more diverse than engine types. Steam generators can use coal, gas, oil, geothermal heat, sunlight or nuclear reactors to heat the water and produce steam. Some steam turbines use geothermal steam direct from the ground.

Gas turbines can use natural gas or man-made gas such as synthesis gas (syngas) made from coal, oil or biomass. They can also use gas from gas-cooled nuclear reactors (see Appendix G.5).

Other engine types tend to be designed to work with a particular primary energy source. Water turbines use water but it could be freshwater from a dam or a running river or seawater from a tidal flow. Wind turbines rely on wind. Petrol and diesel engines, with their own particular fuels, are also used for driving smaller generators that can be started and stopped quickly.

2.5 Power Generator Types

Power generator types are typically identified by their primary energy source. We talk about 'coal power', meaning power from a steam turbine powered by burning coal. 'Hydropower' is power from a water turbine driven by moving water. 'Wind power' is power from a wind turbine driven by moving air or wind. 'Solar (thermal) power' uses heat from the sun to produce the steam that drives a steam turbine.

The one exception is 'nuclear power' which is power from a turbine driven by heat generated by a fission reaction in a nuclear reactor. The primary energy source is actually uranium or thorium but perhaps this is splitting hairs. We will tend to use the term 'fission' for nuclear in this book.

We also talk about power 'stations'. A power station is a factory devoted to making electricity. It might contain several generators but generally they all use the same primary energy source. So a coal power station might contain two or three coal powered generators. Coal power stations are often located close to coal mines to minimise coal transport costs as coal is a bulky, solid fuel and coal power stations use a lot of coal.

There is another type of electricity generator that does not use electromagnetic induction. This is the solar photovoltaic (PV) cell. PV cells use semiconductor[1] material (typically silicon) to convert solar radiation directly into electricity. Solar PV cells are used in solar panels that we might put on our roof to generate electricity from the sun. We will discuss solar PV in more detail in Part II—Renewable Energy.

2.6 Comparing Generators

All power generators produce exactly the same commodity—electricity. An electron from a wind turbine/generator is identical to an electron from a coal plant. We do not get 'green' electrons from wind and 'black' electrons from coal—they all just move indistinguishably in the same conductor.

What we can compare is how reliably each produce energy and how much electrical energy they each produce over an extended period, say 12 months.

As we learnt in Chap. 1, a megawatt (MW) is a measure of how much instantaneous power a generator can produce. This is important but it does not tell us how

[1] A semiconductor has an electrical conductivity between a conductor and an insulator.

Table 2.1 Capacity factors for electricity generators

Energy source	Capacity factor range (%)	Average annual energy output per rated MW (GWh)
Coal and Gas	80–90	7.4
Nuclear	80–90	7.4
Biomass	50–90	6.1
Geothermal	40–95	5.9
Hydropower	25–75	4.4
Solar thermal	25–50	3.3
Wave	25–45	3.1
Wind	20–40	2.6
Tidal	20–35	2.4
Solar PV	10–25	1.5

much electrical energy we will get. That is determined by the length of time we can get power out of the generator over a given period. We might be able to get power out of a coal power station for 80% of the year because for 20% of the time it is either under maintenance or not required. On average, we will probably only get 30% of the power out of a wind farm because the wind does not always blow and the wind speed varies. This brings us to the term 'capacity factor'. This might seem a bit technical for some but it is a critical concept when it comes to comparing the effectiveness of various energy sources.

2.7 Capacity Factor

Capacity factor (or 'load factor' as it is sometimes called) for a generator is usually defined as the ratio of the actual energy output from the generator over a year to the theoretical maximum output it would produce if it operated non-stop at its full capacity for the whole year. So, in the example above the coal power station would be said to have a capacity factor of 80% and the wind farm would have a capacity factor of 30%. This means that we get 80% of the theoretical maximum energy from the coal plant but only 30% of the theoretical maximum from the wind farm (if the wind blew continuously).

Table 2.1 shows typical capacity factors for various electricity generators. The third column shows the average yearly energy produced in gigawatt-hours (GWh) per MW of generator capacity.[2] It is very clear from the table that a

[2] To calculate the electrical energy Power Makers can get from 1 MW of generator power in a full year we multiply the number of MW by the number of hours in a year. There are $24 \times 365 = 8{,}760$ h in a year so at 100% capacity factor we would get 8,760 MWh of electricity from each MW of generator power. At 80% capacity factor every MW of coal power would produce $8{,}760 \times 0.8 = 7{,}008$ MWh of electricity in the year but every MW of wind power would only produce $8{,}760 \times 0.3 = 2{,}628$ MWh. Each MW of wind power is only worth 37.5% of a MW of coal power.

500 MW solar PV farm is very different to a 500 MW coal plant. The coal plant would produce five times as much electricity as the solar PV farm so when we compare generator plants it is important that we look at electrical energy generated (GWh) not power (MW) installed.

We will discuss renewable energy sources in much greater detail in Part II and explain why their capacity factors are often much lower than for coal, gas and nuclear (fission) energy sources.

Chapter 3
Poles and Wires

So far we have discussed what electricity is and how it is made. In this chapter, we will talk about how it gets from the generator to our homes and workplaces. We discuss electricity networks and explain the difference between transmission networks and distribution networks. We then discuss transmission losses and introduce the concept of network control.

3.1 Electricity Networks

Power stations are often located close to primary energy sources. A coal power station will be close to a coal mine, a gas power station will be close to a NG pipeline, a hydroelectric power station will be next to a water reservoir, a wind farm will be in a windy location, perhaps on top of a hill, and a solar plant will be where there is lots of sunshine. One exception is a fission energy plant which does not need to be close to its primary energy source. The reasons for this will become more obvious once we get to Part IV—Fission Energy.

Most of us tend to live in cities, towns, and regional communities that are not close to these primary energy sources so we require a network of wires to transport power from the power stations to where we live and work. Electrical energy travels at close to the speed of light, so even if we live a long way away from the power station the electricity we use in our home is generated almost at the same time as we use it.

These networks do not just have one power station connected to the network, they have a number of different power stations of different types. With an electricity network, we are not just relying on a single power station or a single power source for our electricity which reduces the risk of power failure.

These networks are normally divided into two parts. The "distribution" network which distributes electricity to our individual homes and workplaces and the "transmission" network which transports electricity from the power stations to the

M. Nicholson, *The Power Makers' Challenge*, Green Energy and Technology,
DOI: 10.1007/978-1-4471-2813-7_3, © Springer-Verlag London 2012

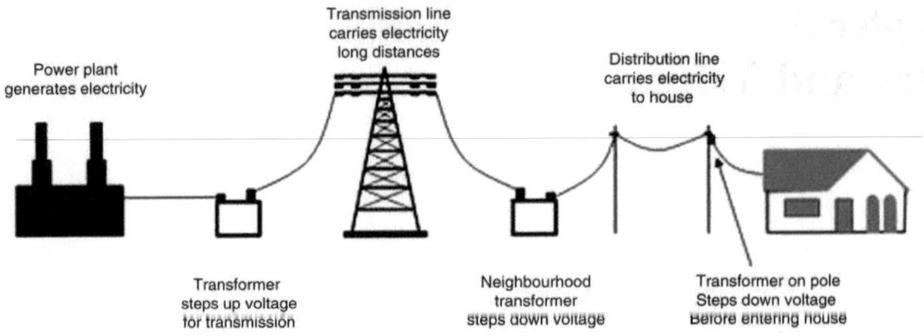

Fig. 3.1 Transmission and distribution network. From US Energy Information Administration (adapted)

distribution network. Sometimes very large industrial customers like aluminum smelters are directly connected to the transmission network. Sometimes some small local generators are directly connected to the distribution network.

Transmission networks usually consist of very tall towers with wires running between them as shown in Fig. 3.1. We have all seen these huge towers and lines snaking across the countryside. But there is also a significant part of the transmission network that is less obvious to the general public. Sometimes the cables are run underground and there are transformers[1] and switching equipment in substations that ensure the network functions effectively.

Distribution networks usually consist of poles and wires, often running along the roadside. The distribution network can also use underground cables as well as substations, transformers, and other equipment.

There is no practical limit to the geographic extent of an electricity network but there are cost and control advantages in having the generators and demand centres evenly distributed throughout the network.

3.2 Transmission Losses

When electricity is transmitted along a wire some of the energy is lost as heat because of what is called "resistance" in the wire. In Chap. 1, the filament in the incandescent lamp used this resistance to generate heat and light and make the torch lamp glow brightly. For the torch, this property of resistance had a positive outcome of providing light. For the transmission and distribution networks this resistance heats the wires and reduces the amount of electrical energy that reaches the end users.

[1] A transformer changes the voltage across the transmission line. It can "step-up" the voltage from a low voltage to a high voltage or "step-down" the voltage. A distribution line might use a voltage of 11,000 V but our home needs a voltage of 240 V. The step-down transformer changes this voltage.

The longer the wire the greater the resistance and the greater the loss of energy. Using thicker diameter wires or more wires reduces the resistance and the loss but there is a limit to how thick you can run a cable economically and more wires means stronger poles and towers.

Another way to address this problem is to transmit the electricity at very high voltage. The voltage used on transmission lines is usually greater than 200,000 V (200 kV). The voltage from the generator might be only 15–25 kV so this has to be increased using a step-up transformer before it is sent over the transmission lines. Closer to our home or workplace the voltage is lowered again using step-down transformers to a final voltage around 100–400 V that we use for our appliances. In practice, several transformers progressively lower the voltage as the line gets closer to the final place of use.

Even using very high voltages still creates transmission and distribution energy losses in the network. These losses can be of the order of 5–10% for some long transmission lines. The losses can be reduced by transmitting the very high voltage as direct current (DC) rather than alternating current (AC), particularly where the cables run under the sea. This is often referred to as high voltage DC (HVDC). There are some losses and costs in converting the generated AC into DC and then back into AC, so HVDC is only used where there are sufficient savings. HVDC typically begins to pay off when above-ground lines are longer than 600 km and undersea cables reach 60 km.

3.3 Controlling the Network

Reliability of the electricity system is a major responsibility of the Power Makers. Most networks set a target for reliability. Some power systems are considered reliable if, over the long-term, they meet at least 99.998% of consumer demand. Another way to express this is that no more than 0.002% of annual energy consumption goes unserved. This is equivalent to about 10 min in a year. The AC frequency (see Chap. 2) and system voltage are key components of that reliability.

If the Power Makers fail to match supply with demand both the frequency and voltage can move away from their normal levels. The frequency must be kept within a very narrow band to protect equipment attached to the network such as our fridges and washing machines. The band usually must be within 1% either side of the normal frequency which is usually 50 or 60 cycles per second depending on where we live.[2]

Similarly, to protect our electrical equipment, the voltage at our home or workplace must not deviate from the normal voltage by more than a few percent.

[2] All directly connected AC networks are effectively a single network using the same AC frequency throughout. This single network will often cover a whole country or may cover several adjacent countries as in Europe.

This requires that all voltages in the network must also not deviate far from their normal value. A power engineer will tell us this is no small task in a large network.

Managing voltage and frequency is referred to as "power quality" and is a key role of the electricity network and is discussed in greater depth in Appendix B—Power Generation and Network Control.

3.4 Advantages of Transmission Networks

The high voltage transmission network gives us access to large and efficient generators that are often located hundreds of kilometers away from where we live. If we did not have that access we would have to rely on generators in our local area which would be smaller and probably generate more expensive electricity because of the economies of scale. We discuss distributed generation later in this book in Chap. 21.

Without transmission networks, we would also need spare capacity in every town and city to ensure a reliable electricity supply involving much greater investment in generator capacity. We will discuss the need for spare capacity in the network in Chap. 4.

Chapter 4
The Balancing Act

So far, we have explained electricity, how it is made and how it is transported. In this chapter we will discuss how we use it and some of the challenges this presents for our Power Makers. We will cover the load curve and its division into baseload, intermediate load and peak load. We then discuss how the Power Makers forecast what the demand will be each hour of each day. Finally, we discuss how the Power Makers maintain reliability of supply through the use of reserve capacity.

4.1 Demand and Supply

Power demand in a network is a measure of how much power we are collectively consuming at a particular moment. There is always some demand in the network. There is always some demand from every home unless the power is turned off at the main switch. We change that demand every time we turn electrical equipment or even a light on or off.

In Chap. 1 we discussed how the electricity supply has to match the electricity demand exactly, moment by moment. In Chap. 3 we mentioned that the Power Makers need to meet our demand with very high reliability. We all hate blackouts, especially when they hit our computers or our air-conditioning at the height of a heatwave. The demand varies throughout the day and changes with the day of the week, the seasons and the temperature so the Power Makers have to continuously alter the electricity supply to follow the change in demand—both up and down.

The electricity demand or 'load' can be forecast 24 h ahead with some accuracy. Figure 4.1 shows a typical load curve demonstrating how the demand in megawatts (MW)[1] varies through the day in a particular network region. Demand is usually at its lowest very early in the morning before we all get up to go to work.

[1] Because demand is the rate at which electrical energy needs to be supplied, it is measured in watts, or multiples of watts, like megawatts (MW) (see Chap. 1).

M. Nicholson, *The Power Makers' Challenge*, Green Energy and Technology,
DOI: 10.1007/978-1-4471-2813-7_4, © Springer-Verlag London 2012

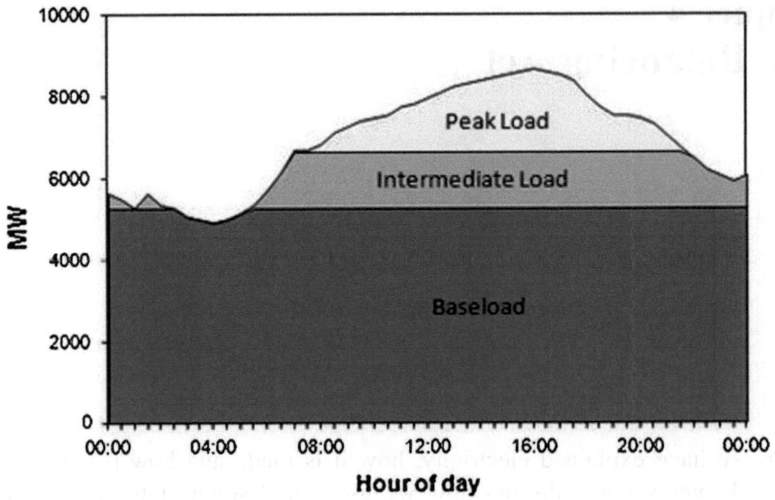

Fig. 4.1 Electricity load curve

As we get up and turn on the heater and electric kettle for tea or coffee and make toast for breakfast the demand starts to rise quite sharply until we leave for work about 08:00 (8 a.m.). The demand then flattens for a while as we all travel to work. It starts to rise again as we start to work and the air-conditioners or heaters kick in at offices and shops. If it is a very hot day the demand will probably peak around 16:00 (4 p.m.) and start to fall through the evening. In winter there will often be two peaks, one mid-morning and the other in the early evening as home heating is turned on.

'Baseload' is the minimum amount of power required to meet the expected customer demand. In Fig. 4.1 baseload is about 5,300 MW. It often represents the bulk of the total demand. Baseload can be around 30–60% of the peak demand depending on the amount of continuously operating industry in the network region and the size of the peak demand (8,700 MW in Fig. 4.1). Baseload power stations typically run continuously to meet this demand throughout the day and night. In most networks, baseload power is supplied by nuclear or coal, gas or oil (fossil fuel) power stations. Some countries use hydropower or geothermal to supply baseload power.

'Intermediate' and 'Peak' load will vary from day to day, largely depending on the weather and whether it is a workday, weekend or holiday. The Power Makers need flexible generators to follow this load as it rises and falls, so they often use gas or oil (diesel) plants and hydropower because these generators are more suited to producing a variable amount of electricity as and when required. The forecast allows these generators to be allocated ahead of time to meet the expected future load and can be what is called 'dispatched'.

We will discuss load variability in much greater depth in Chap. 19.

4.2 Network Reliability

Sometimes things go wrong in a power system. Occasionally generators break down unexpectedly resulting in a significant loss of power to the network. Bad weather or poor maintenance can cause a transmission failure which can interrupt the supply from generators or demand from customers. These events all have an impact on the entire network ranging from very minor to potentially catastrophic. Therefore, all national electricity systems must operate under very tight reliability and power quality standards.

Reliability means that electricity is always there when we need it (see Chap. 3). This requires the network to be able to handle a high demand for electricity and an unexpected loss of a generator *without* impacting supply.

Power quality means that the electricity we receive will operate our electrical appliances safely and not damage them. This requires that any variation in the frequency and voltage in the network must be kept within specified limits as we discussed in the Chap. 3. A significant imbalance between load and supply will impact power quality and may impact reliability.

If generation capacity is lost or the demand rises significantly then non-critical loads can be automatically disconnected from the network to reinstate the balance between demand and supply. This is often called load shedding. If the demand drops significantly over a few seconds (perhaps from a transmission failure) then supply will be reduced by automatically shutting down generators. In the worst case, where the balance cannot be adequately reinstated, it is possible that the entire network will shutdown.

4.3 Reserve Capacity

To avoid such calamities, the Power Makers ensure that adequate generators are scheduled to supply the forecast load *plus* a reserve to meet unexpected conditions. These generators need to respond to slow changes which may happen over a few hours. They must also respond to faster changes in load from second to second so the process has to be automatic. They even need to respond to a sudden and substantial mismatch between generation and load caused by a major fault in the system.

The Power Makers need enough reserve capacity to maintain the reliability and power quality standards in the event of the unexpected loss of the largest individual generating unit or the loss of the most significant transmission line in the network. These reserves operate over different time-scales and together are often referred to as the 'operating' reserve. See Box 4.1 for more information on operating reserves.

Box 4.1 Operating Reserves

What is often called 'spinning' reserve can respond within seconds to increased demand or to address a sudden fault on the network. This requires that these generators are connected to the network but are not running at full capacity (not fully loaded) and can increase their power output quickly to meet the increased demand. Generally these are fossil-fuel plants like coal or gas.

'Standing' reserve is not connected to the network but can be connected and operated within minutes if needed. These reserves need fast response generators and are often supplied by quick starting gas turbines, diesel generators and hydro plants or power imported from another network. Standing reserves may be idle for much of the year.

In addition to the operating reserve, the operator must also maintain what is called a 'reserve margin'. This is the difference between the installed capacity, including power imported from other networks, and the maximum annual peak demand for electricity. This margin is to cover for generators that may be unavailable or being serviced. In today's networks, the operating reserve plus the reserve margin can be 20–30% of the peak demand on the network.

For those interested in greater detail on topics discussed in this chapter then it would also be worth reading Appendix B—Power Generation and Network Control.

Chapter 5
Dollars and Cents

So far we have covered a general overview of electricity networks and some of the responsibilities of the Power Makers. In this chapter we take a look at costs, specifically generating and transmission costs. We introduce the term levelized cost of electricity (LCOE) and discuss the need to get the generator mix right to minimize cost. We then discuss energy efficiency and conservation and their role in reducing the demand for electricity and how they can help a country to meet its energy reduction targets.

5.1 Calculating Generating Costs

It costs money to make electricity. First a power station must be built. Once it is built the station has to be run and maintained. For some power sources like coal and gas, fuel needs to be purchased.

Once the power station is built and operating, then the station can sell electricity to consumers. Power station operators are usually paid a wholesale price for their electricity.

Before investors will invest in a new power station, they need to know the average cost per MWh delivered by the power station over its lifetime. This is derived using a methodology which calculates what is referred to as the "levelized cost of electricity" (LCOE).[1]

The importance of this levelized cost is that it tells investors what the average wholesale price of electricity needs to be to recoup their total investment including capital costs, operating costs, fuel costs, taxes, interest payments, and any other outgoings. Investors of course need to make a profit on their investment so they

[1] The LCOE is usually associated with a specific generating technology. For more details on calculating LCOE see Appendix H—Electricity Costs and Markets and H.1—Levelized Cost of Electricity (LCOE).

M. Nicholson, *The Power Makers' Challenge*, Green Energy and Technology,
DOI: 10.1007/978-1-4471-2813-7_5, © Springer-Verlag London 2012

Table 5.1 Levelized cost for new generation resources, 2016

Energy source	Capacity factor (%)	Levelized cost US $ per MWh
Coal	85	94–108
Coal with CCS	85	135
Gas—CC	87	62–65
Gas—CC with CCS	87	88
Gas—OC	30	100–121
Nuclear	90	113
Biomass	83	111
Geothermal—conversion	90	101
Hydro	51	85
Wind—onshore	34	94
Wind—offshore	39	237
Solar thermal	31	306
Solar PV	21	207

From US EIA [1]

will be looking for an average wholesale price that is higher than the LCOE. If this seems unlikely, then they will not invest.

Wholesale prices vary throughout the day depending on demand and supply. Nighttime baseload power is usually the cheapest as that is when demand is the lowest and the plants typically run continuously. The next most expensive is intermediate load where the demand is reasonably predictable but the power plants used do not run all the time so they need a high wholesale price to recover the investment. The highest wholesale prices are during peak load when demand is highest and supply less readily available and plants run for only relatively short periods.

5.2 Typical Generating Costs

Typical levelized costs for different power station types are shown in Table 5.1. These LCOEs were calculated by the US Energy Information Administration (EIA) [1] for new plants to be installed in the US in 2016. They exclude any transmission cost and carbon costs and are indicative only as they apply specifically to the US.

5.3 Baseload, Intermediate, and Peak Load

We have already discussed different types of demand, baseload, intermediate load, and peak load as shown in Fig. 4.1. The selection of power stations to service different parts of this demand is based on the ability of the power station to supply the load in terms of reliability but also on the cost of producing the electricity.

Investors need to make money. Using a power plant extensively during a period when the wholesale price is likely to be lower than the levelized cost means it may be losing money and it will be more difficult for the investors to recover the initial capital investment.

Baseload power stations typically have high initial capital costs but relatively low running costs. They have high capacity factors (see Chap. 2) which allows them to generate power continuously, reliably, and cheaply. These power stations typically use nuclear power or coal but can also be combined cycle gas turbine (CCGT).

The proportion of demand which varies but is predictable, for example weekday mornings and evenings, is served by what are often referred to as "mid-merit" plants. These power stations generally have lower capital costs and more flexibility to follow the changing load but typically have higher running costs than baseload generators. An example of a mid-merit plant is CCGT.

The peak part of the demand that is highly uncertain, for example, during a hot summer day, is best served by peaking plant such as an open cycle gas turbine (OCGT) or diesel. These power stations have low capital costs and are relatively quick to bring online when needed. They are often part of the reserve capacity we discussed in the last chapter. This flexibility comes at some cost however. They are often required to run inefficiently to allow them to respond instantaneously to sudden increase in demand and this increases their running costs. Hydro plants also provide peak demand in some countries.

5.4 Getting the Mix Right

The best mix of generation is one which minimizes the total cost of meeting demand. The shape of the demand curve on a particular day will influence that mix. For example, a relatively flat demand would mean a greater role for baseload generation having the lowest running cost. A very peaky demand would mean a greater role for peaking generators that can rapidly respond to changes in demand but at higher cost.

The moving wholesale price of electricity can also have an influence on the mix. Whenever the wholesale price is above the immediate costs of operating a particular generator, including fuel and maintenance, then that generator can make a contribution to its fixed costs which are related to the LCOE we discussed earlier in this chapter.

The anticipated return from running the plant, even though the price might be below the LCOE, will determine whether it is economic for the plant operator to enter the market. Ideally, power generators freely bidding into a market should optimize the mix of baseload, mid-merit, and peaking generators.[2]

---

[2] For those that like to skip ahead, we discuss how price affects the mix of generators in greater depth in Chap. 20.

5.5 Transmission Costs

So far in this chapter we have only considered the cost of generating the electricity. As consumers, we need that electricity transported to where we want to use it. The transmission and distribution networks need to be able to handle peak demand at any node in the network. Nodes (or "buses") are connection points throughout the network. As this peak demand grows, the networks need to be upgraded. Peak demand, and not average demand, determines transmission or transport costs.

What is important to consumers is the "delivered cost" of electricity. The delivered cost is the sum of the average electricity wholesale market price (discussed above) and the average cost of transport to the consumer. Typically, about 60% of the delivered cost comes from generation and 40% from transmission and distribution.

Overhead transmission line costs include the cost of towers and lines, the cost of the easements (land used for the towers and lines), and the cost of any substation equipment. The cost is very dependent on the length of the line and the amount of power the line can safely transport. Typical costs for high voltage transmission lines range from $1 to 2 million per km. The cost of underground cabling can be of the order of 10 times the cost of overhead lines. There are also additional costs for the increased repair times for a cable compared with an overhead line.

The Power Makers have a keen interest in these capital costs but as consumers we are really interested in the average cost per MWh to deliver the electrical energy. That is what we have to pay for. The US Energy Information Administration [1] estimated these transmission costs in the US at between 1 and 3% of the LCOE for the power station.

Box 5.1 Example of Transmission Costs

A recent study in Australia [2] to build around 3,000 kms of overhead transmission lines using either AC or HVDC showed capital costs varying from $1.4 to 2.2 million per km for AC and $0.8 to 1.2 million per km for DC. At a transmission load of 5,000 MW the capital costs per watt over this distance ranged from $0.8 to 1.2/W for AC and $0.6 to 0.8/W for DC.

5.6 Energy Efficiency

Let us finish off this chapter on costs by looking at energy efficiency and conservation.

"Energy efficiency" is a rather ambiguous term. It can refer to the energy "conversion" efficiency of a power station which is usually defined as the ratio of the useful energy coming out of the station to the amount of energy going in as fuel to run the engines that drive the turbines.

Alternatively, energy efficiency can also mean "efficient energy use" which means using less energy to achieve the same result.

An example of energy conversion efficiency would be the ratio of the amount of electrical energy that a coal power plant might produce from a unit of coal fuel energy. Typically, the conversion efficiency for a coal plant is around 33% so only a third of the heat energy from the coal will be converted into electrical energy. More modern plants might get over 40% (see Chap. 12).

Efficient energy use for the consumer is about replacing a less efficient device with a more efficient one to achieve the same result. For example, replacing an incandescent lamp with a compact fluorescent lamp (CFL). A CFL needs one quarter of the power of an incandescent lamp to emit the same amount of light so a 25 W CFL will do the same job as a 100 W incandescent lamp and use only one quarter of the electrical energy during operation.

5.7 Energy Conservation

Energy conservation is about reducing the total amount of energy used. It is usually approached by "efficient energy use" as discussed above. Being more efficient with how we do things will reduce energy consumption. Some claim that energy conservation is essential for a move to low-carbon energy. It does have the spin-off benefit of reducing our energy bills and reducing our greenhouse gas emissions. "Negawatthours" (avoided energy consumption) are very popular. Saving energy certainly seems cheaper than making it.

However, there is an economic argument that suggests that if we save money on energy we will spend it on something else. That "something else" will need to be produced using more energy so energy conservation is never actually achieved. History would tend to support this argument as the world energy use per person continues to rise, despite a rapidly developing culture of energy conservation.

Another economic argument is that there is a close correlation between energy use and gross domestic product (GDP). If we were to use less energy then GDP would fall. The developing world is most unlikely to support such a notion so perhaps energy conservation in a global sense is a pious hope. But that does not stop governments of developed nations setting energy reduction targets.

As far as electricity is concerned, population growth will almost certainly guarantee growth in demand for electricity even with conservation. As oil prices rise there will be greater demand for electrified transport which will increase electricity consumption. Where populations are increasing (which is most of the world outside Europe) there will be a need for more freshwater coming from desalinated seawater. Desalination is a big electricity user.

5.8 Energy Reduction Targets

The European Union (EU) has set a target to achieve a 20% energy saving by 2020. This energy saving will come from improved energy efficiency. The EU sees a number of benefits from energy savings including lower greenhouse gas emissions, lower energy bills, and less dependence on imported oil and gas.

The EU believes the opportunity for energy savings are significant. By 2020, they suggest that household energy could be reduced by 30%, energy used in transport could be reduced by 20%, and energy used by industry could be reduced by 19% [3].

Between 1997 and 2006, final energy efficiency in the EU has improved on average by 1.3% per year, so clearly energy efficiency can be achieved, but because of rapid economic (GDP) growth, total energy use in the EU still rose over that decade.

The American Clean Energy and Security Act of 2009 requires new buildings in the US to be 30% more energy efficient by 2012 and 50% more efficient by 2016.

Stopping the growth in energy demand and improved energy efficiency are often seen as vital to achieving a sustainable energy future. Unfortunately, stopping the growth in energy demand probably means stopping the growth in population (attractive to some, anathema to others). The scope for more and more energy efficiency is also limited. We have all heard of the "law" of diminishing returns.

References

1. US DOE/EIA (2011) 2016 Levelized cost of new generation resources in the annual energy outlook 2011. http://eia.doe.gov/oiaf/aeo/electricity_generation.html. Accessed 10 July 2011
2. AEMO (2009) Network extensions to remote areas part 2—Innamincka case study. http://aemo.com.au/planning/0400-0005.pdf. Accessed 10 July 2011
3. EU (2008) Energy efficiency: delivering the 20% target. http://ec.europa.eu/energy/strategies/2008/doc/2008_11_ser2/energy_efficiency_communication_en.pdf. Accessed 10 July 2011

Chapter 6
The Carbon Challenge

The debate about climate change still rages. The Intergovernmental Panel on Climate Change (IPCC) has made assessments of future temperature changes and the reasons for the changes. In this chapter we refresh everyone's memory about climate change and the need to reduce greenhouse gas emissions through a pricing mechanism. We introduce the concept of emissions intensity for electricity generating and look at the differences between the various energy sources already discussed in Chap. 2. We finish this chapter by discussing how the Power Makers will respond to a price on carbon emissions.

6.1 Climate Change Summary

Greenhouse gases are a group of gases in the atmosphere that can trap heat near the Earth's surface. As these trace gases increase in the atmosphere, the extra heat they trap can warm up the Earth in what is commonly referred to as 'global warming'. This warming puts pressure on the Earth's climate system which can lead to climate change.

Carbon dioxide (CO_2) accounts for about 70% of the effects of greenhouse gas emissions released by humans. Carbon dioxide is emitted when fossil fuels such as oil, coal and NG are burned. This could be to make electricity or to run our motor vehicles. It is also released by the clearing and burning of vegetation. Once released, carbon dioxide can have an effective lifetime in the atmosphere of hundreds of years.

Over the last 100 years the average air temperature near the Earth's surface has risen 0.8°C and over the past 30 years the temperature has increased by 0.2°C each decade. The IPCC believes this increase in temperature since the mid-twentieth century is very likely (>90% certainty) due to the greenhouse gas emissions created by us [1].

M. Nicholson, *The Power Makers' Challenge*, Green Energy and Technology,
DOI: 10.1007/978-1-4471-2813-7_6, © Springer-Verlag London 2012

6.2 Reducing Emissions

If we are to slow down, or better still stop, these temperature rises which can cause changes to the climate, we need to reduce or eliminate the amount of greenhouse gases we are releasing into the atmosphere. The IPCC has assessed how much reduction is required for varies target temperature increases.

At the 2009 Climate Summit in Copenhagen it was generally accepted by the world's leaders that the target temperature rise should be no more than 2°C above pre-industrial levels. To achieve this the IPCC tells us we must reduce our emissions by 85% by 2050, worldwide.

Electricity generation is a big contributor to greenhouse gas emissions and will be one of the major target areas for these reductions. According to the World Resources Institute [2], in 2006, 44% of all the world's emitted CO_2 (excluding land use changes) came from electricity generation.

The International Energy Agency (IEA) estimates that world electricity demand will grow at 2.2% per year to 2035 [3]. This means that electricity demand could be more than double by 2050. This growth in demand will make reductions in emissions even more challenging.

6.3 A Price on Carbon

One way to encourage the Power Makers to reduce CO_2 emissions is to charge them for every tonne of CO_2 they emit. This charge is called the 'carbon price'. The price can either be levied as a tax or through the use of emissions permits that the Power Makers would need to buy before being allowed to release the emissions. This carbon price will need to be progressively increased as emission reduction targets are tightened.

6.4 Emission Intensity

On average in the world, about 500 kg of CO_2 equivalent[1] are produced per MWh of electricity generated. This is known as the 'emission intensity'. To reach the emissions reductions needed by 2050 several studies have shown that the average emission intensity needs to be reduced to as low as 50 kg CO_2-e/MWh [4].

The emission intensity of various primary energy sources used to generate electricity can be seen in Table 6.1.[2] The worst offenders are coal and oil, followed

---

1 Greenhouse gas emissions are recorded by their carbon dioxide equivalent impact (CO_2-e).

2 These are life cycle emissions, which means that they cover emissions during power plant construction, fuel mining and transport, operation, decommissioning and waste disposal.

Table 6.1 Emission intensity for different energy sources

Energy source	Emission intensity (kg CO_2-e/MWh)	
	Low	High
Brown coal	1,062	1,372
Black coal	757	1,085
Black coal with CCS	247	
IGCC	795	
IGCC with CCS	130	
Oil	657	866
Natural gas—CCGT	398	499
Natural gas—CCGT with CCS	245	
Nuclear	3	40
Wind—onshore	7	15
Wind—offshore	9	22
Solar PV	13	104
Hydro—reservoir	4	120
Hydro—run-of-river	4	33

CCS carbon capture and storage
IGCC integrated gasification combined cycle
CCGT combined cycle gas turbine
From World Energy Council [5]

by gas. Carbon capture and storage (CCS) will help both coal and gas to reduce CO_2 emissions. We discuss CCS in Part III—Clean Coal. Based on the generation costs shown in Table 5.1, a carbon price of say $30 a tonne of CO_2 could add 35% to the wholesale cost of electricity from coal and 24% to the cost from gas.

The really low emitters are the renewable energy options (like wind, solar and hydro) and nuclear power. We will discuss renewable energy in the next part of this book and nuclear power in Part IV—Fission and Energy.

6.5 The Power Makers' Response

Faced with a price on carbon, Power Makers will look at their generator options. Reducing emissions means either switching primary energy sources or switching to processes like CCS. If these switches can be made more cheaply than paying the price on carbon then they will make the switch. If not, then they will not make the switch unless forced to by legislation. It is as simple as that—they are there to run a business that makes a profit or else close down.

Clearly the Power Makers in most trouble are those running coal plants even if they try to pass the extra carbon cost on to their customers. Electricity is a competitive market in most countries these days. If some Power Makers do not make the switch and others do, then the ones who do not will be disadvantaged by having to charge more for their electricity.

Some power plants can have a 40–50 year life. The Power Makers could be faced with a costly upgrade to a plant, years before they had planned, resulting in significant loss of value in the decommissioned plant. Worse, incorrect technology choices made now could prove to be very expensive mistakes down the track.

This is a testing time for Power Makers. They not only need to make a profit, they also need to provide a reliable service to their customers. Every low-carbon solution for electricity generation has its drawbacks. We will spend much of the time in this book looking at this problem and seeing if we can identify the least cost solution. So let us start with renewable energy.

References

1. IPCC (2007) Contribution of working group I to the fourth assessment report of the intergovernmental panel on climate change. Cambridge University Press, Cambridge, United Kingdom and New York
2. World Resources Institute Climate Analysis Indicators Tool (CAIT) (2011) http://cait.wri.org. Accessed 10 July 2011
3. IEA (2010) World energy outlook 2010. http://worldenergyoutlook.org/. Accessed 10 July 2011
4. Australian Government Treasury (2008) Australia's low pollution future: chapter 5: mitigation scenarios—international results chart 5.22. http://treasury.gov.au/lowpollutionfuture/report/html/05_Chapter5.asp. Accessed 10 July 2011
5. WEC (2004) Comparison of energy systems life cycle assessment. http://worldenergy.org/documents/lca2.pdf. Accessed 10 July 2011

Part II
Renewable Energy

Enthusiasm for renewable energy coupled with a politics in which each nation tries to gain brownie points for its diligence in meeting the Kyoto limits is an unhappy mixture. It will fail and bring discredit both to the greens and to the politicians foolish enough to adopt renewables as a major source of energy before they have been properly developed.

Lovelock J (2006) The Revenge of Gaia
Penguin Books

All renewable energy (RE) sources produce significantly less greenhouse gases than fossil fuel sources when generating electricity. This makes them potentially attractive as solutions to climate change.

We basically have three core RE resources that can be used to make electricity: the sun, gravity, and the Earth's hot core. We will refer to these as renewable resources because they are essentially inexhaustible—they will not run out while ever man lives on the Earth.

The sun is the most powerful energy resource available to us. It provides direct heating and direct radiation, both of which can be turned into electricity. The solar heating of the atmosphere creates winds and waves which can drive electricity generators. The sun also evaporates water to produce rain and snow that can be used for hydro-electricity. Plants grow that can be burned to produce heat or converted into other usable forms (like gas) which can produce electricity. Even our coal came from the sun's ancient radiation because it is derived from plants or other life-forms that lived millions of years ago.

The rate of solar radiation intercepted by the Earth's surface is about 7,000 times the total current rate of world energy consumption, so from that point of view, there is no shortage of solar resources.

Gravitational forces of the sun and moon create the tides that can be used to drive turbines and electricity generators. The Earth's hot core heats water and rock and this heat can also be turned into electricity.

Over the years we have developed technologies that can take these resources and generate significant quantities of electricity. Some of these technologies, like hydro and wind, are mature and widely used. Others like deep hot rocks and wave power are still in development. Continuous improvements are being made to these technologies to improve their usefulness and we can expect many of the current technologies will be producing more and more electricity over the next few decades.

RE technologies will make a growing contribution to the world's energy supply. How big that contribution will be is uncertain and that is what we will explore in this part of the book.

Chapter 7
Many Options

There are many renewable energy (RE) options, so in this chapter we look at the various RE sources and how we turn them into electricity. We discuss the most often proposed RE sources: wind, solar, hydro, biomass, geothermal, tidal and wave. As you will see, some are better at producing reliable electricity than others. We finish the chapter by discussing the low energy density of all renewable resources and the impact this has on amount of resource that has to be captured and the consequential cost imposed.

7.1 Wind

Wind is moving air. The sun heats the Earth's surface unevenly because different types of land and water absorb the sun's heat at different rates. For example, during the day the air above the land heats up more quickly than the air over water. The warmer air expands and rises, and the heavier cooler air rushes in to take its place, creating winds. In a similar way, the high levels of solar energy arriving at the tropics create air turbulence and currents that move around the globe.

Wind turbines use large aerofoil (airfoil)-shaped rotor blades to collect the wind's kinetic[1] energy. The wind energy rotates the blades on a shaft that drives an electricity generator. (See the illustration at the start of Appendix C)

The amount of electricity generated from a wind turbine depends on the speed of the wind. If the wind speed is too low then no power is generated and if the wind speed is too high then the turbine shuts down to avoid damage. In between these extremes the power output increases with speed up to a maximum level. Wind speed increases with the height above the ground so the blades and generator are located on high towers often 100 m tall to capture the higher wind speeds and to support larger rotor blades.

A modern wind turbine can generate over 3 MW of power. They are major structures that, for their size, generate only a relatively small amount of electricity

---

[1] The kinetic energy of a body is the energy it possesses because of its motion.

M. Nicholson, *The Power Makers' Challenge*, Green Energy and Technology,
DOI: 10.1007/978-1-4471-2813-7_7, © Springer-Verlag London 2012

compared to generators in a typical coal power station that will generate 200–600 MW. Commercial wind turbines are usually installed in clusters called wind farms to provide power to the electricity grid. Wind farms can be located on land or in relatively shallow water offshore.

Wind, like most renewable sources, is variable (see Chap. 8). The wind does not blow evenly and sometimes does not blow at all. So the wind speed varies continuously over the day which means the electricity generated from a wind turbine varies as well. Later, we will discuss ways to reduce the impact of this variability.

7.2 Solar

Photovoltaic (PV) cells convert solar radiation directly into electricity. Typical commercial PV cells convert only 10–20% of the solar energy to electrical energy but cells are being developed with higher conversion ratios. PV cells can be used in small clusters such as on a house roof to generate electricity to be used locally or for the grid. The cells can also be installed in large clusters (solar farms) to generate much greater quantities of electricity specifically for the grid.

Solar thermal systems can generate electricity by converting solar energy into thermal energy (as steam) which drives a turbine and generator. To generate the high temperatures needed, thermal systems use sun-tracking concentrating collectors such as parabolic mirrors. The collector focuses the solar energy onto a much smaller absorber which concentrates the energy and heats a fluid such as oil or water. These systems convert 13–25% of the solar energy into electricity. Solar thermal systems for generating electricity are often called concentrated solar power (CSP). Solar thermal electricity systems are nearly always installed in solar farms in dry, hot locations such as deserts. The heated fluid from the absorbers is collected centrally and converted into steam to supply the turbines.

Both solar PV and solar thermal systems need daylight to work and both are adversely affected by cloudy weather. The electricity produced depends on the amount of solar radiation reaching the earth. This changes with the angle of the sun (which varies over the year), the distance from the equator and the amount of cloud cover. This variation is a greater problem for solar PV where a passing cloud will substantially reduce the electricity output. CSP systems have the advantage of being able to store some of the heat to still generate electricity during reduced solar radiation or even darkness.

Like wind, solar power is variable. See Appendix D—Solar Power for more technical details about solar systems.

7.3 Hydro

Hydropower, or power from moving water, is one of the oldest forms of energy. The sun's radiation evaporates water, largely from the sea, and the water vapour rises and cools and eventually condenses into clouds. Some of the rain or snow falls on high ground and flows back to the sea in rivers. Fast flowing water or water that falls from a great height, as in a waterfall, contains a lot of energy. A hydro-electric power plant converts this energy into electricity.

In a run-of-river system the force of the current in the river provides the energy to drive a turbine and generator. In a storage or dam system, the water from a flowing river or lake is collected into a reservoir behind a dam wall and the height of the water produces a fast current to turn turbine blades that drive generators. The amount of electricity that can be generated in a hydro-electric system depends on the availability of flowing water and this can vary considerably throughout the year depending on the flow in the river source, which naturally depends on rainfall or snowmelt.

The storage of river water in a reservoir allows electricity to be generated even at times of reduced river flow—within certain limits. The amount of storage is limited by the size and depth of the reservoir so there may be times during the year when the hydropower station cannot operate. However, hydro is still one of the most available and flexible sources of renewable electricity.

Reservoir water storage can also be used to effectively store electricity as we mentioned in Chap. 1. As we go through the book, we will see that energy storage will be very important for most RE sources. To store electricity in a pumped storage system, water is pushed uphill using electric pumps from a lower reservoir to a higher reservoir during periods of low demand for electricity (off-peak) when it is relatively inexpensive. When the demand increases during peak times and the electricity becomes more expensive, the water is allowed to flow back into the lower reservoir through a turbine to generate electricity. Not all the electricity is recovered in this process, but the price difference between pumping and generating can make it commercially viable.

Hydropower contributes almost 90% of all renewable electricity generated today.

7.4 Biomass

Wood and wood waste are the most common forms of biomass used for electricity generation. The wood can be directly burned to heat water to produce steam for a steam turbine. It can also be heated and turned into gas for use in a gas turbine or an internal combustion engine. Other forms of biomass, such as crop and municipal waste and landfill gas from municipal waste dumps, can all be used as fuels to drive generator engines.

Biomass differs from some other renewable resources like wind or solar radiation in that it can be stored and used when required. However, biomass must be produced and consumed close to the power station because the energy consumed to transport it over long distances could exceed the energy in the fuel itself. Also the land used to grow the biomass is significant (see Chap. 9) and generally means that biomass power plants are relatively small compared to coal or nuclear power plants and are more commonly used for distributed generation where the electricity is used close to where it is generated. Sometimes coal is co-fired with biomass to reduce the coal consumption in traditional coal-fired power stations.

Biomass also differs from other renewable resources in another way. When burned it releases CO_2 into the atmosphere just like coal or gas. So why would we want to use it? Well assuming we specifically grow the biomass for electricity generation, then an equal quantity of the released CO_2 is absorbed by the next generation of growing plants so the net CO_2 going into the atmosphere is zero.

Some biomass fuel can be seasonal such as crop waste and not available all year round.

7.5 Geothermal

Geothermal energy is heat generated from within the Earth. Like hydropower, it has been a source of energy for centuries. The core of the Earth is extremely hot. This heat is mostly generated by the slow decay of radioactive elements. In some places, hot groundwater naturally reaches close to the surface at depths from a few hundred meters to three kilometers.

Conventional geothermal electricity systems use these naturally occurring reservoirs of very hot water and steam to generate electricity using a conventional steam turbine and generator at ground level. This usually requires drilling wells into these high temperature hydrothermal reservoirs which are generally found in regions of volcanic activity. Conventional geothermal systems tend to be limited to a few volcanic regions around the world, such as Iceland, California and New Zealand. Sometimes the reservoir water supply can decline over time reducing the availability of power to the turbines.

Engineered geothermal systems (EGS) are man-made hydrothermal reservoirs that are heated by the surrounding hot rock, often 4–5 km beneath the surface. To create these reservoirs, deep bore holes are drilled down to the hot rock and fluid is pumped down the well under high pressure where it opens existing fractures in the rock (hydraulic fracturing) creating an artificial (engineered) underground reservoir and heat exchanger. To generate electricity, water is pumped down a bore hole through the fractures where it is heated by the surrounding rock. The superheated[2] water is then returned to the surface through another bore hole, to be used to produce steam to drive a generator turbine.

EGS is still in development. Perhaps it ought not to be classified as renewable in any one well because the rate of heat extraction is generally much greater than its replacement from surrounding rock. This means that more wells need to be drilled than are used at any one time but the heat will 'renew' eventually—but it could take a few decades.

---

[2] Superheating is where a liquid is heated under pressure to a temperature higher than its usual boiling point without actually boiling.

7.6 Tidal

Tides are caused by the gravitational pull of the moon and sun and the rotation of the Earth. Near shore, water levels can vary in some places by 12 m or more between tides. The advantage of using tides as a RE source is that, unlike wind and sunshine, they are very predictable (but not constant) and they work day and night all year round. Tidal movements cycle roughly ever half day and the tide levels cycle every 14 days.

Conventional tidal plants involve a dam or barrage across an inlet where the tide flows in and out. Sluice gates on the barrage allow the tidal basin to fill on the incoming high tides and to empty through a turbine system on the outgoing tide.

There are two other ways to harness tidal power by using the tidal stream to power a turbine. One involves mounting vertical axis turbines in a tidal fence across a channel. The other is to use horizontal axis turbines like wind turbines that are located in a strong tidal current.

Tidal systems can be made to be two-way so that electricity can be generated on both the incoming and outgoing tide. The amount of electricity generated depends on the speed of flow of water through the turbine which varies with tide cycle. Power can be generated for 3–6 h during each incoming and outgoing tide depending on the phases of the moon and sun.

7.7 Waves

Waves are caused by the wind blowing over the surface of the ocean. The total power of waves breaking around the world's coastlines is significant but waves are less predictable than tides because they rely on the variable wind.

Wave energy can be harnessed by several methods. One is to direct the waves into a narrow channel to increase their power and size. The waves can then be focused into a reservoir or used directly to spin turbines. Another is to use the mechanical energy from a rising and falling floating body or buoy on the surface of the waves to drive a generator.

The challenges with wave power are to convert the wave motion into electricity efficiently and to build devices that can handle extreme waves, storm damage and corrosion from salt water.

7.8 Dilute Resources

So far we have identified seven separate renewable resources that can be harnessed and turned into electricity. Unfortunately all of these resources have relatively low 'energy density', that is we need to capture a substantial amount of the resource to generate the quantity of electrical energy we need. This has meant huge wind turbines and thousands of them, huge areas of solar collectors and many square kilometers of them, enormous dams to trap water and millions of hectares of land for biomass.

A dilute resource also means much greater cost to extract the energy and convert it into usable and reliable electricity. Technology efficiency measures can go someway to reducing this cost, but there are some physical limits to how far improved efficiency can continue to reduce costs. Developers of potentially scalable[3] renewable technologies (mainly solar and wind) have been predicting major cost reductions for decades. When *properly* costed against reliable coal and nuclear power these technologies still fall short. This does not imply there is no place for renewable electricity generation for some markets but it does indicate that renewable sources will not readily replace significant amounts of coal power which is what we need to reduce GHG emissions.

7.9 Variability

The word 'variable' has cropped up a few times in this chapter. So let us explore what this really means for RE in the Chap. 8.

[3] Capable of being scaled up to replace fossil fuels.

Chapter 8
Nature's Curse

Nature has blessed us with plenty of renewable resources that can be turned into electricity. She then cursed us by making these renewable resources not only very dilute but variable—not there all the time when we need them. The sun does not always shine, the wind does not always blow, or the water does not always flow. In this chapter we look at the natural cycles of each of the RE sources and discuss the impact that variability has on the electricity network. We look at negative loads and forecasting variable resources and finish by discussing some of the constraints of managing RE variability and introduce some possible solutions.

8.1 Natural Variability of RE Resources

All RE resources are variable over different timescales. Table 8.1 shows how these can vary over time from seconds to minutes in the case of solar PV to decades for geothermal power.

Variability in the power generated by RE sources can have significant implications for electricity networks as we will discuss in the next section. This variability can be more easily managed if the variation is over longish timescales as with hydropower, biomass, and geothermal power. Where the variability is over short durations and difficult to predict accurately, as with wind and solar PV, then this short-term variability requires careful management in the electricity network.

As natural cycles vary significantly between different RE sources, combining them in the right way can smooth out some of this variation. Wind and hydropower can be complementary. For example, a country like Norway that gets almost all its power from hydropower can cope with some seasonal variations in hydropower by using pumped storage (see Sect. 7.3) and using wind power when it is available to provide power to the network at peak times and power for water pumping at off-peak times. The wind power and pumped storage reduces the need to use the water in the reservoir or river.

DOI: 10.1007/978-1-4471-2813-7_8, © Springer-Verlag London 2012

Table 8.1 Natural cycles of renewable energy sources

Source	Cycle-time	Main influence	Comments
Geothermal	Decades	Heat availability	Heat and water loss from individual wells
Hydropower	Seasons	Water flow	Seasonal rain, snow pack runoff
Biomass	Harvest cycle	Fuel supply	Depends on source used
Solar thermal	Daily, Hourly	Dark, cloud cover	Heat depends on solar radiation
Solar PV	Daily, Minutes	Dark, cloud cover	Power output depends on brightness
Wind	Daily, Hourly	Wind speed	Power output depends on wind speed
Tidal	12 h/14 days	Ocean flow	Depends on periodic ocean surface flow
Waves	Daily, Hourly	Wind and sea	Depends on wind and fetch

8.2 Why RE Variability is a Problem

Oil, coal, gas, and uranium are available when we need them because these energy resources are easy to store. This means we can generate electricity from them as and when we want it. We cannot store the wind or the tides or sunlight so these sources of energy are not always available when we need them.

The product of these energy resources, electricity, is very expensive to store in large quantities. Large energy systems, like the ones that supply our grids, have little storage capacity so virtually all the electricity we use has to be generated at exactly the same time as we use it, as we discussed in Chap. 1. This requires the Power Makers to continuously balance the generated supply with our demand for electricity.

The only RE sources that can reliably produce power when it is needed are hydropower using a reservoir, pumped storage, biomass, and geothermal, and these are the only RE sources that are normally considered to be dispatchable[1] and able to follow the load. Wind, solar PV, CSP without storage[2] and wave power are not considered dispatchable.

To completely eliminate non-renewable, high greenhouse gas emission energy sources like coal and gas from the electricity networks we would need to use RE sources that can fully replace them or we would have to change the way we meet demand. In reality we will probably attempt to do both but either approach will be very challenging.

8.3 Negative Load

Variable RE generators such as wind and solar PV are sometimes not scheduled, which means they can be run at any time when there is wind or sunlight availability. At these times they are treated as "negative" load. It is called negative load because

[1] See Definitions in Appendix A—Definition of Terms.

[2] CSP with auxiliary gas or sufficient heat storage may become dispatchable but is still vulnerable to lengthy periods of cloud cover.

the output from a wind farm or solar panel reduces the second-by-second demand on the network. When they are generating electricity, these non-scheduled RE sources have the same impact as turning off a piece of electrical equipment—they free up more electricity for other loads on the network. Hence, the term negative load.

During the time that RE sources are acting as negative load they reduce the demand on conventional generators such as coal or gas plants in the same way as turning off the air-conditioning in a building. During these periods they are able to reduce—to some degree—the emissions coming from the coal and gas plants.

Unfortunately, the variations from some RE generators such as wind power can be substantial over relatively short time frames and this can make life more challenging for the Power Makers charged with balancing load with demand. It may require RE energy to be wasted (discarded or curtailed) if they have too much of it in part of the network. Remember that supply must not exceed demand and supply cannot exceed the capacity of the local distribution and transmission system. A significant amount of energy is sometimes discarded in networks with high wind penetration. By discarded, we mean that the wind turbines are disconnected from the network so no electricity is produced.

It can also mean that additional reserve capacity is needed to ensure the system remains reliable and with the right power quality. Sometimes, the transmission and distribution lines need to be strengthened to handle the peak output from the variable RE generators even though that output will often not be available from the generators.

8.4 Forecasting

RE generators, particularly wind and solar PV, present an even greater challenge to the Power Makers if a significant proportion of the total interconnected electricity supply (>20%) is to be provided from these highly variable RE sources. Just as the Power Makers forecast the electricity demand, they also need to forecast RE supply with no guarantee that it will be delivered. For wind and solar, this largely involves weather forecasting which is discussed in Appendix C.8—Wind Forecasting.

Such forecasts are never perfect and significant wind capacity might be lost because of geographically widespread wind calms or solar capacity lost because of cloudy conditions. Some of this variation in supply can be managed with the existing operating reserve (see Box 4.1) but additional system reserve will be required as the penetration of variable sources increases—and this comes at a cost.

8.5 Measuring Variability

We often hear the argument that we can manage variability of solar and wind sources by spreading the generators widely across the state or country. The argument runs that the wind will always be blowing somewhere so if the wind

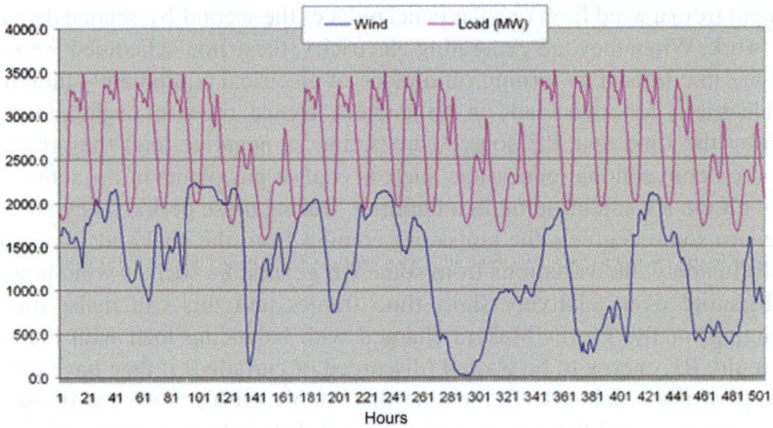

Fig. 8.1 Wind power production and load Western Denmark (Jan 2005). From Energinet.dk (National Danish Transmission System Operator)

farm in one location is becalmed then another in a different location may well be spinning merrily. A similar argument can be made about cloud cover blocking the sunlight.

Indeed there has been several theoretical modeling studies that suggest the output from geographical spread wind farms generally varies smoothly and only rarely experiences a situation where there is no wind at any site [1]. But as more wind is being installed around the world, we are starting to get experiential evidence that this may not be the case.

Denmark has a significant amount of wind in its network and is the largest wind energy producer per capita in the world. It is often quoted as a wonderful example of what can be done with renewable energy. Figure 8.1 shows the total network load over 21 days in West Denmark in the winter month of January. The load naturally goes up and down each day much like in Fig. 4.1. Below the load line is the corresponding total wind production over the same period from all the wind farms in West Denmark (an installed capacity of 2,400 MW).

It is clear the wind energy production in the total network also goes up and down but with much greater irregularity. It varies substantially from day to day. On some days there is close to wind peak production but other days there is almost no wind energy produced at all. There is little correlation between the wind power and the load.

We commonly see similar graphs from other regions around the world and during different periods of the year. Some locations get more wind energy in winter than summer but daily (and even hourly) variations can be substantial at any time of year. A recent study in Canada showed that geographic separation of wind farms did not reduce variability for a given fleet of wind generation [2]. Australia has one of the most geographically spread wind networks anywhere in the world and it produces wind output very similar to Denmark shown in Fig. 8.1.

8.6 Variability Across Generators

RE vendors and supporters will sometimes play down this variability issue with wind and solar by saying that coal and gas plants are also unpredictable because they breakdown sometimes. This is true but misleading because breakdowns happen rarely compared to sudden cloud cover or drop in wind. Coal and gas plants are rarely unpredictable but they do need more scheduled maintenance than solar panels or wind turbines.

Although the wind blows about 80% of the time, the actual amount of electricity generated depends on the wind speed and this varies continuously. Modern wind turbines are designed to reduce the impact of wind speed change by producing a set amount of electricity once the wind reaches around 45 km/h but below this speed the output still varies significantly. Unfortunately, in most places, the wind is not often above 45 km/h (see Fig. C.2 in Appendix C—Wind Power) so variability is a constant problem.

Solar variability is sometimes defended by saying it produces power everyday. Again this is true, but the amount of electricity produced depends on the brightness of the sunlight and is significantly reduced when a cloud blocks the sun. Solar PV output can be as low as 10% of the rated capacity on a day with moderate to heavy cloud cover. Concentrated solar power without thermal storage is even more dependent on direct sunlight and the output might be zero for even light cloud cover.

8.7 Managing Variability

We do have some options for managing short-term RE variability in an electricity network. They include:

- Providing additional conventional reserve and system capacity which comes at a cost
- Storing more electricity
- Interconnecting with other grids that have spare capacity
- Managing demand to better match the load to the RE variations
- Curtailing or discarding some of the RE energy that cannot be used
- Introducing distributed generation.

We will deal with each of these options as we progress through this book but we also have RE technologies that do not suffer from short-term variability so let us look at those in the Chap. 9.

References

1. Sinden G (2005) Characteristics of the UK wind resource: long-term patterns and relationship to electricity demand. Energy Policy J doi:10.1016/j.enpol.2005.10.003
2. Adams T, Cadieux F (2009) Wind power in Ontario: quantifying the benefits of geographic diversity, engineering institute of Canada, 2nd climate change technology conference, May 2009.http://tomadamsenergy.com/wp-content/uploads/2009/05/windpowergeodiversitybenefits_adams_cadieux-colour-graphs-and-citation1.pdf. Accessed 11 July 2011

Chapter 9
Nature's Saviors

Our Power Makers need to provide us with a very reliable electricity supply, 99.998% of the time. Generators that produce highly variable energy output create headaches. What the Power Makers prefer are generators that can produce a predictable amount of energy as and when it is needed—not at the whim of nature's curse. In this chapter we look at the most promising RE technologies that are both reliable and proven. These come down to hydropower, conventional geothermal, biomass and concentrated solar thermal with some thermal storage (CSP).

9.1 Reliable and Proven

When we use the term reliable, we mean the energy production is not variable on short-time scales (seconds to hours). By proven we mean they have demonstrated commercial viability in terms of cost and reliability.

The first three technologies considered to be both reliable and proven are commercially established but may be constrained by the availability of resources; that is water, geothermal hot water (or steam) and combustible biomass. CSP is proven technology but is currently expensive and still only being constructed on a relatively small scale.

Other potentially reliable RE technologies such as engineered geothermal systems (EGS) are not yet proven technologies but we will include them in this chapter as potential saviors for the future.

9.2 More Hydropower

Hydropower is one of the most flexible sources of electricity for the Power Makers. The energy resource (water) can be readily stored in large quantities, which makes it suitable for baseload generation in some countries. Hydro turbines

M. Nicholson, *The Power Makers' Challenge*, Green Energy and Technology,
DOI: 10.1007/978-1-4471-2813-7_9, © Springer-Verlag London 2012

do not rely on heating fuel so they can be started very quickly (within seconds to minutes). This means hydro is also suitable for both intermediate and peak loads. In addition, hydro electricity produces no direct greenhouse gas emissions.

Pumped hydro storage systems (see Sect. 7.3) provide energy storage to assist with balancing the whole network. This is particularly valuable with wind and solar power.

Hydropower (excluding pumped storage) produced about 17% of the world's electricity in 2005 [1]. Although this had fallen from around 22% in 1980, the IEA expects world hydropower to grow by 1.8% a year in real terms to 2030 with the increase being mainly in developing countries [2]. This rate of growth, which also excluded pumped storage, will actual reduce hydro's percentage of total electricity production to 14%.

Unfortunately, most of the low-cost hydroelectric resources have already been exploited in Organisation for Economic Co-operation and Development (OECD) countries and conservationists are not keen on seeing more large dams. But there are still opportunities for adding capacity in the developing world. In the non-OECD nations, hydropower is the predominant source of renewable energy growth, with mid- to large-scale hydro plants expected to be completed in China, India, Brazil and a number of nations in Southeast Asia [3].

New reservoirs present some challenges for the environment and this is a significant limiting factor when it comes building new large hydro plants. However, according to the World Energy Council (WEC) there are 45,000 large dams in the world and the majority do not have hydro plants [4]. It may be possible to build new hydroelectric plants on some of these dams without significant impact on the environment.

It is also possible to increase the output from existing hydro plants by upgrading to improve efficiency with new turbines and generators. This is generally the least cost way of boosting hydro energy output.

Other proven hydro solutions like run-of-river systems do not need dams so the environmental impact is much less. The hydro turbines are generally much smaller than with dam systems and they are vulnerable to the river flow which can vary in the medium-term (days-months). The size and scope of run-of-river hydro means that it can only make a small contribution to decarbonising our electricity. They are probably more applicable to distributed generation systems (see Chap. 3).

Modern societies need lots of fresh water so we need reservoirs to store the intermittent rainfall or snowmelt and spread the availability of water over the whole year. Using more of these water storages to generate electricity could certainly assist in moving us away from fossil fuels but it is unlikely to be a total solution.

9.3 More Conventional Geothermal

Conventional geothermal generators that use hot water or steam emitted from deep in the ground generally have a regular supply of steam to drive steam turbines. They are extremely reliable and have capacity factors similar to coal plants.

This means they can provide the Power Makers with baseload and intermediate load supply just like coal. Geothermal is not dependent on weather patterns unlike hydro which can suffer from droughts. It also requires a relatively small land area compared to hydro, biomass and solar thermal.

Conventional geothermal resources are widely distributed throughout the world, but only 24 countries have temperatures close to the surface that are high enough for generating electricity [5]. Only 0.3% of world electricity comes from these conventional geothermal resources so it is not a big player in the scheme of things. However, several countries, such as Iceland and the Philippines, which are in the lucky 24, get more that 15% of their electricity from conventional geothermal resources [4].

There are clearly opportunities for many of those 24 countries, such as Indonesia, to increase their use of geothermal electricity. But for the rest of the world, they will need to look elsewhere to replace their fossil fuels.

9.4 More Biomass

Biomass (largely wood or wood waste but also some crop and municipal waste) can be burned to heat water to generate steam. Some biomass is co-fired with coal in conventional coal plants. Biomass can also be turned into syngas for use in a gas turbine. So biomass is really a coal or NG substitute for the Power Makers but without the potentially costly net CO_2 emissions (see Chap. 6).

Electricity from biomass represents 1% of world electricity so again it is a relatively small contributor. So, if biomass is a good coal substitute with no net carbon emissions (see Sect. 7.4), why would not we use more of it? The problem lies in the land needed to grow the wood.

A 500 MW biomass power station would need about 2 million tonnes of dry wood each year. Based on some design proposals, fast growing trees will need 175,000 ha—an area of land 42 by 42 km—to supply a 500 MW power station with an ongoing 20-year supply of fuel [6]. This land would need to be fertile and dedicated to trees so it may compete with food crops. It would also need to be close to the power station to minimise transport cost and pollution. In practise, biomass is generally used in small to medium-sized generators.

A better option might be to use more municipal solid waste (MSW)—trash and garbage that currently goes to landfill and recycling. Up to 90% of this solid waste by volume can be burned to produce electricity—often referred to as waste-to-energy. This still produces CO_2 but if the waste was buried in landfill it would just rot and release methane which is an even more potent greenhouse gas than CO_2.

So how much of our electricity could realistically come from MSW? Mackay in his book *Sustainable Energy—without the hot air*, estimated that the average person in the United Kingdom (UK) produces 517 kg of MSW each year [7]. Assuming that incinerating 1 kg of waste yields roughly 0.5 kWh of electricity then a four person household could get 2–3 kWh per day from MSW. Given that many households of 4 people use 20 kWh of electricity per day then MSW could provide 10% of their electricity. This would assume that very little MSW was recycled.

Assuming we continue to throw out trash and garbage, there is clearly an opportunity to replace some fossil fuels with MSW but it will not really address the carbon emissions problem unless we see landfill as part of the natural carbon cycle.

9.5 CSP with Storage

Solar thermal technology has one big advantage over the other highly variable RE technologies. It produces heat, which is more readily stored than electricity in order to eliminate short-term variations in energy output. There are several proven technologies for doing this and these are explained in some detail in Appendix D—Solar Power.

Technically, it is possible to put enough storage on the system to allow it to run continuously throughout the day and night, and even over several days of cloudy weather. This allows the power makers to use CSP in the same way they would use a coal plant for baseload and intermediate load. But it all comes at a significant cost. Particularly, when the CSP plant has to work as well in winter as it does in summer. In some locations, the difference in available solar energy between summer and winter can often be 4:1 (see Appendix D—Solar Power).

More heat storage means more hours of operation for the plant, but it also means a larger solar field of solar collectors. This is the most expensive part of the plant so a big solar field substantially increases the construction cost and land use. A 500 MW plant capable of continuous operation could need over 3,000 ha (30 km^2) of land. This is significantly greater than for geothermal but significantly less than for biomass and even less than some coal plants including the mine sites (see Table 11.3).

The plant needs to be installed on flat land preferably in the solar belt between 20 and 40° north or south of the equator.[1] This rules out a large part of the world. A European proposed solution to this problem is to instal CSP plants in the Sahara Desert in Africa and then transport the energy 2,000 km north. This adds another cost of course. Transmission links are not cheap, as we discussed in Chap. 5.

For some users, a more attractive option may be to use CSP for intermediate and peak load only. This use would require a smaller amount of heat storage (an hour or two) with a smaller area of solar collectors. It could only be used on sunny days or use gas as backup to extend the use on cloudy days. This option does not eliminate fossil fuels completely but it could reduce the greenhouse gases produced.

---

[1] Areas closer to the equator are generally excluded because of high rainfall and associated cloud cover.

As we also saw from Table 5.1, solar thermal is one of the most expensive solutions today for electricity generation but the CSP industry expects costs to come down over time as more plants get built (see Sect. 7.8).

9.6 EGS

Engineered geothermal systems (EGS) might offer the Power Makers the same service as conventional geothermal systems. The difference between the two is below the ground not above it. EGS aims to create similar conditions that occur naturally with conventional geothermal where a steady stream of steam, heated by the earth below the power plant, is available for the turbine. The process is briefly explained in Sect. 7.5.

The key attraction for EGS is that heat is naturally present everywhere in the earth at depths from 3 to 10 km so it is available to everyone in theory. However, the heat of the rocks varies widely, which limits the scope for EGS in practise. Despite substantial worldwide potential and considerable effort around the world over the last 40 years, we are yet to see our first commercial-sized production system. The technology to drill and fracture rock has been used extensively in the oil and gas industry for many years but EGS presents many challenges to the would-be EGS plant builder. The tricky part is in creating and sustaining the EGS reservoir.

The US Department of Energy (DOE) performed an analysis in 2008 on EGS and identified 44 separate task steps in the path towards producing electricity [8]. At the current stage of development only 5 of these were considered to be zero risk. That left 39 with some level of risk that the task could not be completed successfully. The DOE identified several technological advancements that will clearly be needed to ensure economic success. They went on to say that specific technology requirements at this stage of EGS reservoir development remain uncertain due to limited operating experience.

EGS is yet to be demonstrated on a commercial scale so is still high risk technology—albeit very promising.

9.7 Summary

So there are some RE solutions that can be seen as potential replacements for some fossil-fuel-powered generators. The acute observer will have noticed that there is no mention of wind, wave or solar PV in this chapter. Without energy storage, these solutions will only make a minimal contribution to replacing existing fossil fuel plants. They will reduce the load on the network and possibly delay the need to instal more coal or gas plants, but they will not add significantly to capacity reliability. In the next chapter let us look at what large-scale energy storage solutions are available.

References

1. US DOE/EIA (2006) World net electricity generation by type (billion kWh) 2005. http://www.eia.doe.gov/iea/elec.html. Accessed 12 July 2011
2. IEA (2009) World energy outlook 2009. http://worldenergyoutlook.org/. Accessed 10 July 2011
3. US DOE/EIA (2009) International energy outlook 2009. http://eia.doe.gov/oiaf/ieo/index.html. Accessed 24 Jan 2010
4. WEC (2007) Survey of energy resources 2007. http://worldenergy.org/publications/survey_of_energy_resources_2007/default.asp. Accessed 12 July 2011
5. WEC (2009) Survey of energy resources interim update 2009. http://www.worldenergy.org/publications/survey_of_energy_resources_interim_update_2009/default.asp. Accessed 12 July 2011
6. Ragland K, Ostlie L, Berg D (2000) WTE biomass power plant in central Wisconsin. http://doa.state.wi.us/docs_view2.asp?docid=54. Accessed 12 July 2011
7. MacKay D (2008) Sustainable energy—without the hot air. UIT Cambridge, Cambridge
8. US DOE (2008) An evaluation of enhanced geothermal systems technology. http://www1.eere.energy.gov/geothermal/pdfs/evaluation_egs_Tech_2008.pdf. Accessed 12 July 2011

Chapter 10
Storing Electricity

In this chapter we cover the all-important issue of storing electricity. It may be the one thing that can make renewable energy a major contributor to reliable electricity. We first discuss the need for energy storage. We then cover the most common method of bulk storing electricity which is pumped-hydro followed by the next most promising which is compressed air energy storage or CAES for short. We discuss two possible future bulk storage systems for electricity which are hydrogen and chemical batteries. We finish this chapter by looking at storage costs and a possible replacement for storage which is demand management.

10.1 Why Do We Need Energy Storage?

The Achilles' heel of variable renewable energy is the inability to guarantee uninterrupted power supply. Unless we can fix this, variable RE sources like wind and solar PV will not be able to replace coal and gas as baseload generators.

A solution is energy storage to smooth out the variability over the day. The storage would allow excess electrical energy (above the needs of the network at the time of generation) to be stored in some form and later released during periods of reduced supply—during wind lulls or cloudy days and nighttime for example.

Storing electricity in the network is not new. Relatively small-scale storage systems like flywheels and superconducting magnetic energy storage already assist Power Makers to improve the reliability of the network through short-term power balancing, voltage regulation, and frequency control (see Appendix B—Power Generation and Network control).

If we were to rely on variable RE sources for the majority of our electricity we would need to handle variations in demand differently. The minute-by-minute variance could be managed with flywheels and batteries. Variations on a timescale of 30–60 min would need pumped storage, compressed air energy storage, (CAES) or hydrogen with some possible contribution from batteries. Longer term variations between 1 and 3 h would still need much larger storage systems.

M. Nicholson, *The Power Makers' Challenge*, Green Energy and Technology,
DOI: 10.1007/978-1-4471-2813-7_10, © Springer-Verlag London 2012

The variation challenge is in the >30 min timescale which needs bulk storage to deliver the power levels and energy required quickly. We have already discussed one such bulk storage system in Chap. 7 which is pumped storage so let us review it more closely.

10.2 Pumped-Hydro Storage

Pumped storage has been used for almost as long as electricity networks have existed. It is the largest capacity energy storage system in use today and represents over 99% of all electricity storage. Pumping typically takes place at night or weekends when there is surplus low-cost electricity to run the pumps and lift the water from the lower reservoir to the higher storage reservoir (see Sect. E.3 in Appendix E—Energy Storage).

When electricity is needed from the storage then the water flows back to the lower reservoir through the hydro turbines. It typically services peak load demand but also provides standing reserve capacity and frequency control services.

More recently, with the increased use of wind power, pumped-hydro storage (PHS) is providing storage services to wind by pumping to the storage reservoir during periods of excess wind supply and providing backup services during periods of reduced wind supply.

The financial benefit of PHS systems is to buy cheap electricity to pump the water up hill and generate more expensive electricity during peak demand. PHS systems are expensive to build but have a long life over which to recover the construction and operating costs. However, in the UK, it is still cheaper to use standing reserve such as open cycle gas turbines (OCGT) than PHS [1].

There is over 100 GW of PHS generating capacity installed worldwide. Although this represents about 4% of the world generating capacity this actually does not tell us very much. What we really need to know is the amount of electricity that can be generated from the stored water. Most PHS systems use reversible turbine/generators that work as both pumps and generators which means they cannot pump and generate electricity at the same time. So they can never be generating for more than half the time however much water storage they have. In practise, installed PHS systems have a capacity factor closer to 15%—in other words they can generate electricity for only about 4 h a day on average. Assuming a 15% capacity factor then 100 GW of PHS could produce less than 1% of world electricity consumption.

If a network without any interconnections to neighbors elected to get 10% of its electricity from variable RE sources it might need much more storage capacity than just 1% of its total consumption to rely solely on the storage to handle the variability. If the 10% were all wind power it might need more than the full 10% depending on how many consecutive days the network experienced little or no wind (see Chap. 8). If the network had a daily demand of 1,000 GWh (about 85% of the UK), 10% would mean a storage of 100 GWh.

There are obviously environmental issues with PHS systems. They need large areas of land for the reservoirs (although the sea could be the lower reservoir) and the upper reservoir needs to be as high as possible, usually a head of more than 200 m above the lower reservoir. The smaller the head the larger the reservoirs need to be. These geographical requirements and the need for two reservoirs limit the number of suitable locations and environmental issues will need to be considered.

If we wanted to get all of our electricity from renewable energy, we might need at least one full day's storage to handle variability. Storing 1,000 GWh would mean storing 2,000 GL of water all with a head of 200 m (see Sect. E.3 in Appendix E—Energy Storage). To give us an idea of what 2,000 GL of storage means, Melbourne in Australia has a population of just under 4 million and a total water storage capacity of 1,800 GL. So our 2,000 GL would be sufficient water to supply a city of 4 million people. But of course you could not use the water for both water supply and pumped storage at the same time. Water supply storage is intended to go up and down with the seasons. We need our PHS system to be full all the time. If the terrain does not allow a 200 m head but only 100 m then we would need to double the volume of water stored.

The largest PHS system in the UK is Dinorwig in Wales which stores only 9 GWh. So we may need to split up our 1,000 GWh of storage across 100 sites. What is worse we would actually need twice the storage area because we need to hold the water in the lower reservoir so it can be pumped back up to the upper reservoir. Unless we can use the sea of course but that means finding high cliffs close to the coast and building a salt water reservoir at the top of the cliffs. None of this will sound very attractive to conservationists.

PHS clearly has its place in modern electricity networks. That is why it is widely used when suitable sites are available. But given the limits on its capacity to scale up, it is not likely to be a panacea for large-scale variable renewable energy.

Later in this Chapter we will briefly look at storage costs per MWh. PHS, despite its high construction costs, proves to be by far the cheapest electricity storage option over its lifetime of any other practical electricity storage media.

10.3 Compressed Air Energy Storage

Another mature bulk storage technology, although not widely used, involves storing compressed air in underground mines or caverns (see Sect. E.4 in Appendix E—Energy Storage). The stored compressed air can later be used to drive a turbine/generator. There are only two CAES systems in operation; one in Germany which began operations in 1978 and can store about 0.6 GWh, and one in the US which can store 2.6 GWh, but others are planned. To store our hypothetical 1,000 GWh we would need almost 400 plants of similar size to the existing US plant.

CAES systems can be used in much the same way as PHS systems but they need much less surface land area and they do not need access to large volumes of water. CAES systems are also less expensive to build than PHS systems. But current generation CAES systems do use some gas so they are not free of fossil fuels. It is possible that this will change in the future.

CAES requires suitable geological sites that already contain disused mines, gas/oil fields or aquifers that can be used as compressed air storage areas. Alternatively, formations like salt domes where storage caverns can be created might be suitable. Such sites will not be available everywhere but globally they are not uncommon.

Where such sites are available in a suitable location for a wind farm then a Wind/CAES combination power plant could be made dispatchable and supply baseload power. It would even be possible to use a wind turbine to drive the compressor directly without the intermediate electricity generation process.

10.4 Hydrogen Energy Storage

For those without sufficient suitable storage sites for compressed air, perhaps they could store hydrogen. Hydrogen needs less than one hundredth of the storage volume of compressed air to store the same amount of energy, depending on the storage pressures (see Table E.1). Hydrogen does not occur naturally on the Earth. To use hydrogen for stationary electricity storage it needs to be extracted from water using a process called electrolysis (see Sect. E.5 in Appendix E—Energy Storage).

Although the smaller storage volume is attractive, the electrolyzers and fuel cells needed for hydrogen energy storage and release are not cheap and are likely to make hydrogen storage significantly more expensive than using compressed air. Hydrogen is also difficult to compress and store safely and efficiently.

A few small-scale demonstration plants using wind turbines and fuel cells and/or hydrogen engines have been built over the last decade but no 'bulk' scale plants have yet been commissioned. For more technical details about hydrogen energy storage (see Sect. E.5 in Appendix E—Energy Storage).

10.5 Batteries

Batteries come in many shapes and sizes and are generally self contained storage devices. They are usually very expensive systems on a cost-per-unit-of-energy-stored basis. In terms of bulk energy storage they tend to be at the small end of town and are often used for distributed generating systems such as relatively small wind or solar plants under 50 MW providing up to 10 MWh of electricity storage.

Table 10.1 Energy storage costs per cycle

Per cycle cost	US $ per MWh Low	US $ per MWh High
Pumped-hydro	1	15
CAES	20	50
Flywheel	30	220
NaS battery	80	250
Flow battery	50	800
Lithium-ion battery	150	1,000
Nickel–cadmium battery	200	1,000
Lead-acid battery	200	1,000

From Electricity Storage Association [2]

The batteries we are all most familiar with are lead-acid batteries—the type used to start our cars. They have traditionally been used for electricity storage and backup systems in telephone exchanges for decades. They tend to be bulky and new and cheaper battery technologies are being developed.

Sodium-sulfur batteries have been commercially available for 20 years and there are several installations under 10 MWh, although systems have been scaled up to a few hundred MWh. Remember we could be looking for several hundred GWh (a GWh is 1,000 times bigger than a MWh), so this would require a lot of batteries.

Lithium-ion batteries are a developing technology showing good promise for electric vehicles as well as stationary energy applications. Again these are generally relatively small storage devices. A Li-ion battery pack for an electric vehicle will store about 20 kWh. We would need 50,000 such battery packs to store just 1 GWh for the grid.

Flow batteries operate a bit differently from conventional batteries because they store electricity by using chemicals stored in large tanks. At the present time they tend to be low-powered and used for smaller storage systems between 5 kWh and 2 MWh. They are still very expensive and are best suited as output levellers for smaller wind or solar installations. Appendix E discusses batteries in more detail.

10.6 Energy Storage Costs

Table 10.1 shows the cost-per-cycle of storage for various storage media we have discussed so far. This table gives us the cost to charge and discharge the storage medium per MWh taking into account the expected life (number of "cycles") and the round trip efficiency of the storage medium. The cost only includes the capital cost but this will be the major cost component for any of these media.

The high cost of batteries compared to pumped-hydro or CAES explains why batteries have not been popular for bulk energy storage except where they provide the only practical alternative as is the case with electric transport (see Chap. 23).

No cost data were available for hydrogen storage.

10.7 Demand Management: An Alternative to Storage?

Demand management aims to match demand on the network to the available supply. It is possible for the network to automatically disconnect certain loads (machinery, appliances, etc.) if the electricity supply falls to certain levels. This can have a similar result as an energy storage system. Instead of providing additional supply from the storage, demand management reduces the demand to bring the network back into balance. In other words, some users are disconnected from the network to keep it in balance. For a discussion on how this is done, see Sect. B.6 in Appendix B—Power Generation and Network Control.

Clearly not all loads are suitable for demand management but appliances like water heaters, refrigerators, pool pumps, and air conditioners can be disconnected for short periods (15–20 min) without impacting the end-user. Although we need a refrigerator to always maintain a constant temperature, it does not need to draw power continuously to achieve this. In fact refrigerators already turn themselves on and off in cycles of the order of 5 min to 1 h so the refrigerator can make its off cycle to be at times of high demand on the grid and defer the load until the demand reduces. Appliances like smart dishwashers or clothes driers can also be scheduled to be used in known periods of low demand.

Such demand management systems can be readily implemented in industrial and commercial settings where demand is generally more predictable. It will be more difficult and costly to implement with domestic customers and may require us to make lifestyle changes.

Demand management has some distinct advantages over energy storage in that it involves smaller capital costs and no efficiency losses. However, it does require sufficient suitable loads to be disconnected when needed without the end-users being unreasonably disadvantaged.

10.8 Energy Storage Summary

It is clear from the above that we really only have a few limited, scalable options for electricity storage. This is a field that has attracted a great deal of interest over the past few decades. Improvements in battery technology are progressing and developments are taking place in both CAES and hydrogen storage. But there are some chemical and physical limitations that will restrict future size and cost reductions.

The most promising (and cheapest) storage technologies are still pumped-hydro and CAES but both require appropriate geological sites. Of the geologically independent technologies hydrogen may deliver but despite several decades of research no commercial scale electricity storage plant using hydrogen is yet in operation.

There are significant economic issues involved in electricity storage as it involves high capital costs and operating complexity in energy markets [1]. Demand management using smart networks may be able to make a contribution but it is still early days. We look at smart networks later in Chap. 22.

References

1. UK Parliamentary Office of Science and Technology (2008) Electricity storage, post note 306. http://parliament.uk/documents/upload/postpn306.pdf. Accessed 14 July 2011
2. Electricity Storage Association (2009) Technology comparisons per cycle cost. http://electricitystorage.org/ESA/technologies/. Accessed 14 July 2011

Chapter 11
Smoke and Mirrors

Many questionable comparisons are made between renewable energy (RE) sources and conventional energy sources like coal, gas, and nuclear. Both sides of the debate can be guilty of some "sleight of hand" in these comparisons. In this chapter, we look at a few of these sleights along with a few other myths that seem to plaque the renewable energy debate. We take the opportunity in this chapter to introduce the term "capacity credit" and to discuss the raw material needs and land use of various energy options used to generate electricity. We finish by discussing the baseload myth from both sides of the debate.

11.1 Comparing Apples with Oranges

It is natural to try to make comparisons between different energy sources. But comparing wind and solar to coal and gas can be like comparing apples with oranges. Just as apples and oranges are both fruits and the wind and coal are both energy resources used to produce electricity—there the comparison ends.

The major difference is that wind and solar photovoltaic (PV) only produce electricity as the energy is collected, whereas both coal and gas can be collected or mined at any time and then readily stored for later use. This is a huge difference. The Power Makers have no way of storing the wind or the sunlight so we only get electricity when the wind is blowing or the sun is shining. Assuming they have collected and stored sufficient coal and gas ahead of use then these resources are always available 24 h a day so they can make electricity when we want it.

Wind and solar energy vary continuously. The amount of electrical energy we get from a wind turbine or a solar PV cell is directly related to wind speed and solar radiation, so as the wind and solar energy varies so does the electricity output. We can forecast to some extent whether it is likely to be windy or sunny but we cannot forecast how the wind will actually blow at any moment or where the clouds will be that mask the sunlight, yet these are the factors that control the energy output.

M. Nicholson, *The Power Makers' Challenge*, Green Energy and Technology,
DOI: 10.1007/978-1-4471-2813-7_11, © Springer-Verlag London 2012

In comparison, stored resources like coal, gas, and oil, and the renewable storable resources like water and biomass can produce a regular electricity output that is readily controlled. As we saw in Chap. 4, the Power Makers need to continuously match supply with demand. It is not difficult to see why they would use coal or gas rather than wind or solar to meet our demand for continuous high quality power.

An example of a bizarre comparison is when wind and solar system suppliers describe the amount of energy their system will produce by saying how many homes it will power. There is a misleading implication that somehow those homes could rely on this system alone. This analogy might make sense for a gas generator that could indeed reliably supply those homes with electricity as and when the home owners needed it. This can never be said for wind or solar PV, which can only ever provide part of the energy needed unless the system came with adequate electricity storage to provide the energy during the times that the wind and sun could not.

11.2 Comparing Energy Outputs

Remember in Chap. 2 we talked about comparing generator types and how a megawatt (MW) of wind power is not the same as a MW of gas power? A MW is a measure of how much instantaneous power a generator can produce. This is important but it does not tell us how much electrical energy we will get, or whether it is dispatchable. That is determined by the capacity factor and reliability, as we covered in Chap. 2.

From Table 2.1 we see that wind has a capacity factor of 20–40% and gas has a capacity factor of 80–90%. Solar PV panels might only have a capacity factor of 10–25%. This means that we get around 3 times as much energy from a MW of gas than a MW of wind and 5 times as much than from a MW of solar PV.

This difference between using units of energy rather than units of power or capacity can have a big impact on many comparisons discussed in this chapter.

11.3 Capacity Credit

What is important to our Power Makers is not just how much electricity a generator could produce in a year, but how much contribution can be made to network reliability (see Chap. 4). All generators can break down occasionally and they may not be able to produce the electricity when it is needed. The "capacity credit" or "capacity value" of a generator is the amount of additional demand or load that can be serviced by that generator at the required reliability level. It is a measure of the contribution that a generator makes to system reliability. Another way to express it is the amount of "guaranteed" capacity from a generator that does not impact reliability. This is sometimes called the "firm" capacity. It is important that we do not confuse capacity credit with capacity factor.

The capacity credit for a CCGT plant is likely to be close to the capacity factor. During scheduled maintenance we would not expect any power from the generator and this is included in assessing the capacity factor. The chance that a gas generator would breakdown unexpectedly might be between 1 and 10% depending on the technology and the age of the plant. A simple way to compare the two is:

Capacity factor ⇒ It will be there sometimes (that percentage of the year)
Capacity credit ⇒ It will be there when you need it (that percentage of the capacity)

For variable RE sources the reliability issue is not so much about unexpected mechanical breakdown but the expected sudden loss of the energy source—the wind or solar radiation. Within any one network, this depends on the seasons, the number of RE plants (wind or solar PV farms) and the geographical distribution of those variable plants. Geographical distribution can spread the risk of low wind or cloud across a wider area and reduce that risk but not eliminate it—as we saw in Chap. 8.

With a relatively small amount of RE sources the capacity credit can be close to the capacity factor, but as the penetration increases, the capacity credit falls because the total network impact from the loss of the wind or sun will be that much greater.

The capacity credit for a variable RE source like wind and solar can be defined as the amount of conventional fossil-fuel plant that can be replaced and maintain the same system reliability. For wind, the capacity credit will fall from the capacity factor (which might be 30% of the installed wind capacity) to 20% or even to below 10% as the amount of installed wind capacity increases. These figures come from theoretical analysis of high penetrations of wind power in various networks [1]. Not everyone agrees with these figures, however. See Sect. C.9 in Appendix C—Wind Power for a more comprehensive review of wind capacity credit and the impact on network system reserve capacity.

The US Department of Energy (DOE) in a recent report titled "20% Wind Energy by 2030" [2] states that wind generators have limited capacity value and that wind should be seen as an energy resource not as capacity resource. The DOE report states that we should not think in terms of "backing up" the wind because the wind capacity was installed to generate low-emissions energy not to meet load growth requirements. By implication, if wind power is not adding to generating capacity, then it will not replace existing fossil-fuel plants (which has typically been the experience in the world so far). Hopefully, it may reduce emissions by reducing the energy coming from fossil-fuel plants, but it is unlikely to actually close any plants down.

11.4 Comparing RE Penetration

The concept of RE penetration in a network presents another comparison trap for the unsuspecting. Some like to talk about RE penetration as meaning the amount of RE power capacity installed in MW as a percentage of the peak power load on the network. This is comparing apples with oranges again because it assumes that a

MW of wind or solar power is the same as a MW of gas or coal power, which it clearly is not (see above).

A much more meaningful way to express RE penetration is to use a unit of measure that is the same for all generators. A MWh of electricity is the same whether it comes from wind or gas. So a far better way to express RE penetration is as the percentage of the total electrical energy generated in the network over a year—known as the "energy penetration". At least this way we are comparing like with like and there is a significant difference between the two methods in terms of percentage penetration.

A 20% capacity penetration (the percentage of peak load) is likely to be only a 10% energy penetration (the percentage of electricity generated). This means that penetration numbers often overstate the true case in terms of energy generated.

11.5 Comparing Different Networks

Even when the RE penetration trap is avoided, there is a tendency to compare networks without regard to where that network is and what it might be connected to.

Both West Denmark and North Germany have significant wind energy penetration; 24% for West Denmark and 33% for North Germany [1]. We often hear exclamations that if these networks can function happily with these relatively high wind energy penetrations why cannot our networks do the same? The answer is they probably can if they have strong network interconnections between your network and one or more of your neighbors and a neighbor can supply power from non-wind sources when you need it to cover periods when there is no wind.

West Denmark has strong high voltage direct current (HVDC) interconnections with Norway and Sweden that can carry more power than the wind can generate in West Denmark. Both Norway and Sweden have vast hydro resources, so West Denmark is able to buy power from its neighbors to cover its entire wind capacity if needed. As we saw in Chap. 8, sometimes they might need it. North Germany has interconnections with neighbors that have twice the capacity of their wind power.

Many countries and regions do not, or cannot, have such accommodating neighbors and they have to limit the amount of variable RE generation in their network.

11.6 Comparing the Costs of Electricity

As we saw in Chap. 5, there are number of parts to calculating the cost of generating electricity from a particular resource such as gas or solar. They typically include the cost of building the generator plant, the cost of operating and maintaining the plant, the cost of the fuel used by the generator and the cost of the money invested (largely the interest paid). These costs need to be considered over the expected lifetime of the plant which might vary from 10 to 50 years or more depending on the technology.

Just comparing the construction costs per MW can be very misleading. The cost of building and operating the plants can vary significantly. Hydro, nuclear, and solar plants tend to be very expensive to build per MW compared to coal or gas plants. But the fuel costs for hydro and solar plants are zero whereas the fuel cost for gas plants over their lifetime will be significantly more than the construction and operating costs. The fuel costs for nuclear are not zero but they are very small compared to gas.

We looked at the levelized cost of generation of several plants in Table 5.1. These costs take no account of external costs such as the environmental cost or the costs of powerlines to integrate them into the electricity network.

Powerline costs are particularly relevant for renewable energy projects that often need to be located some distance from the main load centers or cities. Solar thermal plants are likely to be built in desert regions hundreds of kilometers from where the electricity will be used. Wind farms with good winds often need to be in hilly rural areas well away from the end customers for the electricity and the powerlines need to be able to handle the peak output from the wind farm, not the average power. Generator costs like those shown in Table 5.1 do not generally include the cost of transmission, so costs of solar thermal, geothermal, and wind sources may need to take this into account before comparing them with fossil-fuel or nuclear plants built close to cities.

From Table 5.1 it is clear that solar is the most expensive way of generating electricity today. This is partly due to low capacity factors for solar plants (see Table 2.1) and the low conversion efficiency in turning sunlight into electricity, but the construction cost per MW is also high. This is expected to improve over time but there are some physical limits, in capturing and converting sunlight, that cannot be overcome.

11.7 RE Energy Use

Some critics of renewable energy have argued that RE sources use more energy to make them than they ever give back. Table 11.1 clearly shows that this is not the case. The table contains various assessments of the number of months that the RE systems have to operate to recover the energy (and compensate for the associated greenhouse gas emissions) that went into creating the system. Note this is *not* the same as the financial payback period. The financial payback period is the number of months (or, more likely, years) that the system has to operate to recover the cost of the system and will be much longer than the energy payback period.

With energy payback periods of less than 4 years and generator lifetimes of greater than 10 years, it is clearly wrong to suggest that any of these RE sources use more energy to make them than they can produce over their lifetime. Note that these values are highly dependent on the location of the RE sources and the methodology used in the calculations and are included just to give an indication of possible payback periods, rather than give firm numbers.

Table 11.1 Energy payback period

Energy source	Months	Comments
Wind—onshore [3]	2–3	Danish wind industry analysis
Wind—offshore [4]	13	UK Govt—independent analysis
Marine [5]	20	Carbon trust assessment
Solar PV—Rooftop [6]	12–48	Depends on technology used—see Appendix D—Solar Power
Biomass [4]	42	UK Govt—depends on biomass source
Hydropower [7]	22	University of Sydney—ISA

11.8 Comparing Efficiencies

It is sometimes claimed that very little power reaches the end user from a centralized coal power plant. This is based on the notion that the conversion efficiency is very low for a coal power plant and power is lost in the transmission lines as we discussed in Chap. 3. In one sense this is true. In most coal plants only a third of the energy in the coal is converted to electricity. Another 10% could be lost in the transport so only 30% of the primary energy in the coal reaches our homes as electricity.

Unfortunately, solar power is much worse. A modern commercial solar PV module will convert solar energy to electricity with an efficiency of about 14% (it can range from 10 to 20% depending on the module). So only 14% of the absorbed light landing on the module will be converted to electricity. The inverter that converts the direct current from the PV module to alternating current for the grid is about 90% efficient so only 12–13% of the energy from the sun gets into the grid.

CSP has a conversion efficiency from sunlight to electricity of around 15%. This includes the conversion from solar radiation to heat by the concentrating collectors and the conversion from heat to electricity in the steam turbine.

11.9 Comparison of Building Materials

All power plants need to use building materials, mostly steel and concrete. Nuclear power is often criticized as a huge user of materials. This is true if you just look at the materials used to build a power station without considering the amount of energy the power station generates over its life. Nuclear power stations tend to be very high energy producers so you would expect them to be large and need a lot of building materials. Building materials are often quoted in tonnes per MW (power plant size) rather than tonnes per MWh (power plant energy generation). This can mislead us into thinking that nuclear power uses more resources than solar panels when the opposite is true.

Table 11.2 shows the concrete and steel used in some plant constructions expressed as tonnes per GWh per year. The capacity factors shown were the ones

Table 11.2 Building materials

Material (t/GWh/year)	Capacity factor (%)	Concrete	Steel
Nuclear	85	43	8
Solar PV farm	20	43	10
Wind—onshore	30	159	43
Solar thermal—7.5 h storage	44	338	105

From ISA [7] except solar thermal from NEEDS [8]

used in the referenced reports. Of these plants, nuclear power has the lowest use of concrete and steel per unit of energy generated in one year. If the full lifetime of the plants had been considered then the nuclear plant use of concrete and steel would be even less compared to the other technologies because nuclear plants have some of the longest life spans. A nuclear plant will last at least 40 years but a solar panel may last less than 20 years.

11.10 Comparing Land Use

Various advocates of a particular technology will often use land use as an argument against competing technologies. With some technologies like wind there is the risk of apple and orange comparisons. Do you count the whole area of the wind farm in the calculation or just the footprint of the wind turbine plus access roads? The difference between the two can be a factor of 50. The area between the turbines can still be used for grazing cattle (but not growing trees for biomass). See the picture at the start of Appendix C—Wind Power. Similarly, with coal plants where some coal mines use surface strip mining and others use underground mining. You can still use the land above an underground mine for farming.

To avoid further apple and orange mix-ups, Table 11.3 below identifies the land use per GWh per year to ensure a like comparison between the technologies. GWh output depends on capacity factors so these are included in the table. Again these are the capacity factors used in the particular reference material. In practice of course, solar will not achieve 100% capacity factor so the land use per GWh will be higher than the figures shown. Hydro is a special case because the land use depends on the head height, the average reservoir depth and the flow rate through the reservoir but this is a typical number.

Nuclear is the most land efficient (although this doesn't include the uranium mining but that is not significant per GWh[2]) and biomass is looking like a big challenge except for a country with plenty of spare arable land. Solar thermal is often criticized for requiring vast tracks of land but it is still more land efficient than surface strip coal mining according to Massachusetts Institute of Technology (MIT) based on the questionable 100% capacity factor.

---

[2] A nuclear power plant uses about 0.002 m^3 of ore per GWh.

Table 11.3 Comparing land use

Energy source	Capacity factor (%)	Land use (m^2/GWh/year)	Comments
Nuclear	90	450	Plant site including cooling water
Geothermal	95	900	Flash plant including wells and pipes
Wind—onshore	30	1,100	Turbine footprint plus access only
Solar thermal	100	3,200	Desert based—6 h storage
Coal (strip mining)	80	5,700	Including mining site
Solar PV	100	7,500	Solar farm with dedicated land
Hydro (reservoir)	100	200,000	100 m head, 20 m depth
Biomass	86	460,000	Tree area with 20 year fuel supply

From MIT [9] except wind from NREL [10], Hydro and nuclear (author calculation), biomass [11]

For those who like to tread gently on the earth, look for solutions closer to the top of the table.

11.11 Baseload Myths

Let us finish this chapter on smoke and mirror comparisons by discussing what is sometimes called the baseload "myth". It usually refers to claims from the fossil-fuel and nuclear industries that renewable energy sources "can't do baseload", but as usual, there are two sides to every story.

Baseload usually represents the bulk of the total electricity demand (see Chap. 4). It is largely serviced by coal plants which are the largest emitters of greenhouse gases so we need to find new sources of baseload power if we are to reduce these emissions. Many see renewable energy as the natural, clean, and safe replacement choice.

There are renewable energy sources that can and do provide baseload power in some countries. These are usually restricted to hydro, conventional geothermal, and biomass. There are limited supplies of these resources in many countries as we discussed in Chap. 9 so they would not be able to replace all coal plants on their own.

Unfortunately, the natural variations in other renewable energy resources, particularly wind and solar, make it difficult to dispatch plants that rely on them. It means they cannot reliably make a contribution to baseload without substantial energy storage or addition reserve capacity using fossil fuels like gas. As discussed earlier, some countries like Denmark have access to neighboring networks which allow them to buy electricity from elsewhere when the wind stops blowing in their own country. This allows them to satisfy some of their baseload requirement with a wind/energy-import combination.

Some advocates of using much more of these variable resources for baseload have argued that all power stations are variable to some degree—coal and gas plants breakdown sometimes and need scheduled maintenance—and the electricity network seems to cope with these outages. All wind or solar power is doing, they argue, is adding a bit more variability.

But as we discussed in Sect. 11.3 Capacity Credit earlier in this chapter, reliability is key for Power Makers and the issue of generator variability and hence reliability is incorporated in calculating the capacity credit of a single generator. Coal and gas plants have much higher capacity credits and can make a much greater contribution to network reliability than variable RE generators.

Some solar advocates have suggested that we should move loads from night to day to reduce the need to supply as much baseload power. This would also bring the load into the time of day that suits solar best. If that were straightforward and simple to do it could have some merit. But most of the nighttime load is used by 24 h operating industries like metal smelting and assembly lines so time shifting loads would be impossible for these industries.

Urban society is becoming more round-the-clock with late night shopping and evening sports events. We also enjoy more evening entertainment watching high energy plasma screens or home entertainment systems. Before long, more and more of us will be buying plug-in electric vehicles (PEV) (see Chap. 23) that we will want to recharge in the garage as we sleep at night. The load demand of these vehicles could double our domestic consumption of electricity. The Power Makers must service the needs of their customers, not dictate when they can have power and when they cannot.

Some even argued that there is no such thing as "baseload". The argument runs that all plants together supply the network, and the network serves all loads and no particular plant needs to run continuously to serve a particular load so we should stop building any more baseload plants [12].

Unfortunately this argument ignores the Power Makers' duty to maintain power quality as well as quantity. Variations in voltage and frequency must be kept within narrow bounds to avoid loss of service (leading to loss of production) or damage to electrical equipment. This is much more manageable with a high proportion of the generating capacity providing a steady, reliable output (reliable generators supplying baseload) and then managing the demand variations above the baseload by bringing additional capacity on line as it is needed (see Chap. 4). It seems unlikely that the existing baseload need will disappear anytime soon.

Charles Bayless, a former US utility company CEO, provided an excellent description of why baseload generators are essential for system reliability in an article titled "The Case for Baseload", published in 2010 [13].

Pretending baseload does not exist or could be reduced will not get rid of the need for coal plants. We need a practical technology fix for coal plant emissions, if it exists, and that is the topic of Part III—Clean Goal.

References

1. IEA Wind (2007) Task 25 Design and operation of power systems with large amounts of wind power. http://ieawind.org/AnnexXXV/Publications/W82.pdf. Accessed 19 July 2011
2. US DOE (2008) 20% Wind energy by 2030 Report. http://20percentwind.org/20p.aspx?page=Report. Accessed 19 July 2011

3. Wind Power Note (1997) The energy balance of modern wind turbines No. 16, Dec 1997. http://apere.org/manager/docnum/doc/doc1249_971216_wind.fiche37.pdf. Accessed 19 July 2011
4. UK Parliament (2004) Select committee on science and technology fourth report, appendix 8: energy payback times. http://publications.parliament.uk/pa/ld200304/ldselect/ldsctech/126/12620.htm. Accessed 19 July 2011
5. Carbon Trust (2011) Life-cycle energy and emissions of marine energy devices. http://carbontrust.co.uk/SiteCollectionDocuments/Various/Emerging%20technologies/Technology%20Directory/Marine/Other%20topics/Life-cycle%20energy%20and%20emissions%20of%20devices%20.pdf. Accessed 19 July 2011
6. US DOE (2004) PV FAQS, What is the energy payback for PV? http://nrel.gov/docs/fy04osti/35489.pdf. Accessed 21 July 2011
7. ISA, University of Sydney (2006) Life-cycle energy balance and greenhouse gas emissions of nuclear energy in Australia. http://isa.org.usyd.edu.au/publications/documents/ISA_Nuclear_Report.pdf. Accessed 21 July 2011
8. NEEDS (2008) Final report on technical data, costs, and life cycle inventories of solar thermal power plants. http://needs-project.org/docs/results/RS1a/RS1a%20D12.2%20Final%20report%20concentrating%20solar%20thermal%20power%20plants.pdf. Accessed 21 July 2011
9. MIT (2006) Future of geothermal energy impact of Enhanced Geothermal Systems (EGS) on the United States in the 21st century. http://geothermal.inel.gov/publications/future_of_geothermal_energy.pdf. Accessed 21 July 2011
10. NREL (2011) Wind farm area calculator. http://nrel.gov/analysis/power_databook/calc_wind.php. Accessed 21 July 2011
11. Ragland K, Ostlie L, Berg D (2000) WTE Biomass power plant in central Wisconsin. http://doa.state.wi.us/docs_view2.asp?docid=54. Accessed 12 July 2011
12. Lovins A (2009) Nuclear nonsense. http://rmi.org/rmi/Library/2009-10_NuclearNonsense. Accessed 21 July 2011
13. Bayless C (2010) The case for baseload, electric perspectives. http://eei.org/magazine/EEI%20Electric%20Perspectives%20Article%20Listing/2010-09-01-BASELOAD.pdf. Accessed 21 July 2011

Part III
Clean Coal

Coal remains indispensable. No other fuel matches its promise of cheap and abundant energy for development.

Victor D, Morse R (2009) Living with Coal, Boston Review

Coal is the major source of the world's electricity. It is relatively easy to mine, transport, store, and convert into electricity and many countries have their own coal resources. This makes it the cheapest source of electricity. It is a tough act to follow.

What may kill coal is the emissions. But many countries will not give up their cheap coal easily. Hence the search for "clean coal"—advanced coal technologies that can deal with the emissions problem.

Chapter 12
Why Clean Coal?

Today, coal is indispensable in most electricity networks. Over 40% of the world's electricity is generated in coal plants. Coal is arguably the cheapest way to make electricity and coal is liberally distributed around the planet. Proven reserves worldwide will last more than 120 years [1]. That is the good news for coal fans. The bad news is coal has the highest emission intensity of all energy sources as can be seen from Table 6.1. It is also the most damaging energy generation source locally, from soot, ash, sulfur, and other fine particles that are released in vast quantities from the smokestack, causing chronic health problems among exposed populations. Despite producing 42% of the electricity it generates 75% of the CO_2 emissions coming from electricity generation. If we are to reduce electricity emissions we have to fix up coal. So how do we do it? That is what we discuss in this chapter. We look at ways to reduce coal emissions by improving conversion efficiency and carbon capture and storage (CCS).

12.1 Reducing the Coal Emission Intensity

To some people "clean coal" is an oxymoron. Anyone who has held a piece of coal knows how dirty it is, so how can it be clean? Well it can never be clean in that sense but it can be less polluting.

We have a couple of ways of reducing the polluting emissions from coal. The first is to use less of it to generate the same amount of electricity. In other words to improve the conversion efficiency (see Chap. 5). The second is to capture the emissions before they escape to the atmosphere and then store them somewhere safe. Capturing and storing carbon emissions (often referred to as carbon capture and storage) is worthy of a whole section on its own so we will deal with it later.

M. Nicholson, *The Power Makers' Challenge*, Green Energy and Technology, DOI: 10.1007/978-1-4471-2813-7_12, © Springer-Verlag London 2012

Let us consider how we can improve the conversion efficiency of coal. In fact we have been doing that for many years. A hundred years ago, when solid fuel was used in the boilers, efficiencies were generally below 30%.[1] In other words less than 30% of the stored energy in the coal was converted into electricity. In the 1920s they started pulverizing the coal into a fine powder before putting it into the furnace which increased the combustion temperature and conversion efficiency to around 35%. In the 1950s, the temperature and pressure of the steam in the boiler was raised to what is known as "supercritical" to improve the efficiency even further to 40%. Many coal plants in operation today use pulverized coal and supercritical temperatures. The most efficient plants are "ultra-supercritical" and run at even higher temperatures and pressures and can achieve efficiencies of 45%. Replacing old subcritical coal plants with advanced ultra-supercritical plants can improve efficiencies by up to 10% points, saving hundreds of thousands of tonnes of coal.

To improve the efficiency beyond 50% requires converting the coal to a gas and using a combined-cycle gas turbine in what is called integrated gasification combined cycle (IGCC). Efficiencies of close to 60% may be possible but this is relatively new technology (only five plants have been built so far) and such efficiencies are yet to be demonstrated.

But even if we improved all our coal plant efficiencies by 50%, coal plants will still have an emission intensity of more than 600 kg CO_2-e/MWh, so it will not get us close to lowering the emission intensity to levels discussed in Chap. 6 which are less than one-tenth of this value. We either need to stop using coal or capture and store the emissions.

12.2 Carbon Capture and Storage

CCS is a set of technologies that can significantly reduce carbon dioxide emissions from new and existing coal and gas power plants.

CCS requires the CO_2 to be captured at the power plant and transported to a suitable long-term storage space. The storage would generally be underground in depleted oil or gas reservoirs, discarded coal beds, or well-capped aquifers (see Appendix F—Carbon Capture and Storage for more details).

Capturing the CO_2 is a costly and energy intensive process. For a pulverized coal plant it can use as much as 40% of the electricity generated by the plant. On cost alone, it seems unlikely that CCS will be implemented within the next two decades without financial assistance above and beyond the likely carbon price [2]. In other words, without government subsidies, it will be cheaper for the Power Makers to continue releasing CO_2 than invest in costly CCS technology.

The storage sites probably need to be within 300 km of the plant to reduce the cost of transportation which will generally be high-pressure pipeline. This means

[1] While this might seem low, the science of energy conversion (thermodynamics) makes it clear that such a large loss of fuel energy as heat is only to be expected.

that not all coal plants will have suitable sites—and therefore be candidates for CCS. Even where suitable sites are available, there is the issue of long-term liability for the security of the CO_2 storage once the site is closed. A future leakage from the site will, at best, undo the reduced emissions and, at worst if the leakage is substantial, suffocate local animals and possibly people. This risk alone may restrict the implementation of CCS.

As we can see from Table 6.1, if we convert a coal plant to CCS the emission intensity is significantly reduced from around 900 kg CO_2-e/MWh to less than 300 kg CO_2-e/MWh. This means reducing the emissions from coal plants by two-thirds. Unfortunately, even this significant reduction is inadequate to make coal a long-term solution for baseload electricity supply on its own. In Chap. 6 we observed that, to achieve the required emission reduction target, the emission intensity of world electricity generation needs to be as low as 50 kg CO_2-e/MWh by 2050. This will not be possible if half the electricity is coming from pulverized coal with CCS.

IGCC with CCS could reduce the emission intensity to under 150 kg CO_2-e/MWh. For those countries that insist on continuing to use coal this technology could be a solution, if combined with other technologies like renewable energy.

A final comment on CCS is on its long-term acceptability. It is really only a developing technology today. Some small-scale pilot plants have been built but it has not been scaled up to work with large commercial generating plants. There are also some question marks around the location and adequacy of storage areas [3] but these issues are dismissed by others [4]. On cost alone, CCS will struggle to gain wide acceptance without a significant price on carbon. After all, it cannot, by definition, ever be made cheaper than coal alone.

12.3 Alternatives to Coal

If we cannot rely on cleaning up all the coal then what else can we use instead? As coal is primarily a baseload generator fuel we need cleaner solutions that can deliver baseload supply. This is the topic of the next chapter.

References

1. World Coal Institute (2011) http://worldcoal.org/. Accessed 21 July 2011
2. US DOE (2010) Report of the interagency task force on carbon capture and storage. http://www.fe.doe.gov/programs/sequestration/ccstf/CCSTaskForceReport2010.pdf. Accessed 21 July 2011
3. Oil Online (2009) Carbon storage just doesn't hold water. http://oilonline.com/default.asp?id=326&nid=16341&name=Carbon+storage+just+doesn%E2%80%99t+hold+water. Accessed 21 July 2011
4. NRDC Switchboard (2010) Economides (x2) try their hand at CCS—and get it wrong. http://switchboard.nrdc.org/blogs/gperidas/economides_x2_try_their_hand_a.html. Accessed 21 July 2011

Chapter 13
Baseload Alternatives

Today, most baseload demand (see Sect. 4.1) is met by coal plants. If we are to replace coal plants we need technologies that can deliver baseload supply. We start this chapter by looking at what makes a baseload power station. We then evaluate what technologies (both renewable and non-renewable) meet the requirements of a baseload power station. We then look at whether gas can really address the need for low-emission electricity. Finally, we introduce the possibility of more nuclear power.

13.1 What Makes a Baseload Power Station?

In Chap. 4 we defined baseload as the minimum amount of power required to meet the expected customer demand. We went on to say that baseload power stations typically run continuously to meet this demand throughout the day and night.

For a power station to run continuously it needs to meet certain criteria. The US government's Energy Information Administration (EIA) suggests the first criterion which is "facilities that operate at annual utilisation rates of 70% or higher" [1]. For baseload plants utilisation rates can be considered to be the same thing as capacity factor so we are looking for plants with capacity factors of 70% or higher. Baseload plants also need to have high reliability, with low forced outage rates (FOR).[1]

From Table 5.1 (also from the EIA) we see that coal, combined cycle gas, nuclear, geothermal, and biomass meet these criteria. The first three are considered to be non-renewable and the last two renewable sources of energy. No other plant on EIA's list meets these criteria. In some countries like Norway and Canada with access to plenty of rain and snow melt, hydro plants can have high capacity factors and low FORs and can be used as baseload plants, but these are the exception.

[1] FOR is the probability that the plant will not be available for service when required.

M. Nicholson, *The Power Makers' Challenge*, Green Energy and Technology,
DOI: 10.1007/978-1-4471-2813-7_13, © Springer-Verlag London 2012

13.2 Which Energy Sources Can Replace Coal?

Capacity factor is very important, but it is not sufficient to define energy sources that can realistically replace significant numbers of coal plants. As well as being able to run continuously on any one day they need to be able to produce electricity in substantial quantities and be cost competitive, to allow coal plants to be decommissioned.

In Chap. 9 we looked at both biomass and conventional geothermal as two of Nature's Saviors. As explained in that chapter, both these sources are resource constrained. Biomass is constrained by the sheer amount of land needed to grow trees for a large biomass plant to run throughout the year—a minimum requirement to replace most coal plants. Conventional geothermal is significantly exploited already in most of the lucky 24 countries blessed with hot near-surface reservoirs, so the scope for further coal replacement with conventional geothermal is limited.

Although not considered by EIA, solar thermal (CSP) with substantial thermal storage and/or gas co-firing might be a candidate in some solar belt countries, but at significant additional cost today. Based on the numbers in Table 5.1, electricity from solar thermal is two and half times the price of coal. Engineered geothermal systems (EGS) may also be a candidate—if we can ever get EGS to work on a commercial scale.

That leaves us with two alternatives today to replace coal in most countries. Combined cycle gas and nuclear power.

13.3 Can Gas Deliver Us From Coal?

As a fuel technology, gas is fine. It burns much more cleanly than coal and produces half to two-thirds of the emissions. NG is less evenly distributed around the world than coal and there are probably only half the reserves of gas. For those countries with energy security concerns and limited access to gas in their own country, coal can look more attractive than gas.

Fuel cost is a significant contributor to the total cost of running a gas turbine. As oil and gas reserves run down this will increase gas prices and increase the cost of the electricity produced.

Combined cycle gas plants (CCGT) have an emission intensity of around 400–500 kg CO_2-e/MWh. This means CCGT cannot replace coal and get us to the ultimate emissions reduction target of 50 kg CO_2-e/MWh. Using CCS with CCGT will help but because of the power loss with CCS the emissions will at best halve using current technology.

Some countries have already started what is called the "Dash for Gas" to replace dirty coal plants with "clean" gas. Others will follow. They consider that gas will get them on the track to reducing emissions from electricity generation. Eventually, they will inevitably have to look for alternatives if we are to make the 2050 emissions reduction target we discussed in Chap. 6.

That brings us to nuclear power. Nuclear is a strong baseload candidate to replace coal as it has in France over the last 40 years. It also happens to have an emissions intensity well under the 50 kg CO_2-e/MWh—the goal we are aiming for in 2050.

13.4 But Isn't Nuclear Dangerous?

There is widespread anxiety and doubt around nuclear power. Some of it is very understandable but some of it misplaced and based on inaccurate information. Next, in Part IV—Fission Energy, we will look at nuclear power (or "fission energy" as I prefer to call it) and tackle these anxieties head on.

Reference

1. US DOE/EIA (2011) 2016 Levelized cost of new generation resources in the annual energy outlook 2011. http://eia.doe.gov/oiaf/aeo/electricity_generation.html. Accessed 10 July 2011

Part IV
Fission Energy

I believe nuclear power is the only source of energy that will satisfy our demands and yet not be a hazard to Gaia and interfere with its capacity to sustain a comfortable climate and atmospheric composition.

Lovelock J (2006) The Revenge of Gaia
Penguin Books

Fission energy was a massive breakthrough in 1951, yet it has hardly been exploited to a fraction of its potential. Since those early days of "atomic" energy, as it used to be called, it has steadily expanded despite some heavy set backs in the 1980s. Unlike other energy sources, it is on the brink of improving its efficiency 100-fold. That is, it will soon be able to get 100 times as much energy out of the same primary resource. This is unlikely to be possible for any existing renewable energy resources or fossil fuels. My guess is that our grandchildren will wonder what all the fuss around nuclear power was really about. Climate change will be the catalyst that changes the equation as the renowned environmentalist James Lovelock courageously predicted.

Chapter 14
Fission, Not Combustion

In this chapter we compare using fission energy (nuclear power) to coal energy for generating electricity. We start by recapping how they both produce electricity. We then briefly discuss how fission reactors work (there is more comprehensive explanation in Appendix G—Nuclear Power) and compare fuel and waste from both fission and coal plants. We finish by comparing the CO_2 emissions and cost for both fuel sources.

14.1 How Do Coal and Fission Energy Produce Electricity?

The answer to this question is actually quite straightforward, and we have already discussed it briefly in Chap. 2. They typically both do it in exactly the same way—by boiling water. The heated water turns to steam in a high pressure steam generator and the high pressure steam is used to drive a steam turbine/generator combination to produce electricity.

The big difference between coal and fission is that coal is combusted (that is, burned in a chemical reaction with oxygen) to boil the water whereas fission relies on a nuclear reaction by splitting uranium atoms. The nuclear fission reaction creates heat as burning coal creates heat and in both cases the released heat is used to boil water.

14.2 Fission Reactors

The most common type of fission reactors are thermal reactors called "light-water" reactors (LWR). Light-water just means regular pure water.[1] There are actually several types of thermal reactors but they all work on much the same principle. An atom (usually a uranium atom) is split apart releasing large amounts

[1] It's called light-water to distinguish it from so-called "heavy water", which contains the heavy form of the hydrogen atom called deuterium.

M. Nicholson, *The Power Makers' Challenge*, Green Energy and Technology, DOI: 10.1007/978-1-4471-2813-7_14, © Springer-Verlag London 2012

of energy in the form of heat which is then transferred to the water. For those who would like to know more about nuclear fission and how it works, take a look at Appendix G—Nuclear Power.

Thermal reactors were first used commercially to generate electricity in the late 1950s and there are now over 400 thermal reactors installed in more than 30 countries worldwide. Together they generate about 16% of the world's electricity. France is one of the largest users of fission energy and gets almost 80% of its electricity from its 59 nuclear power stations.

For comparison, there are probably over 2,000 coal power plants worldwide producing just over 40% of the world's electricity.

14.3 Fuel and Waste

Both coal and fission plants use fuels mined from the earth. Coal for coal plants and uranium for fission plants. The big difference is the amount of fuel that needs to be used. A 1,000 MW coal power station needs about 3–4 million tonnes of coal a year. A 1,000 MW fission plant needs only about 150–200 tonnes of natural uranium a year. That means a fission plant uses only a tiny fraction (1/20,000) of the fuel of a coal plant. It also means that the fission plant does not need to be close to the uranium mine because only a relatively small amount of uranium fuel needs to be transported.

Less fuel used means less fuel to store and less leftover waste. No huge coal storage areas and waste slag heaps containing toxic metals like arsenic and lead are needed for fission plants, and no need for thousands of kilometers of coal freight trains.

Fission fuel is significantly less expensive than coal per unit of energy generated—even though the natural uranium removed from the ground has to be processed before it can be used in a reactor. Fuel in a fission plant makes up about 5–10% of the operating cost of the plant. For a coal plant that can be 30–60%. This means fission energy costs are less vulnerable to increases in fuel prices than coal (or gas) energy.

Current generation thermal reactors actually extract less than 1% of the usable energy from the natural uranium. Coal plants extract nearly all the energy from the coal although they cannot convert it all into electricity.

Coal and fission are both improving their efficiency in process technology. As we discussed in Chap. 12, coal conversion efficiency has improved by 50% over the last 70 years and there is a possibility of it improving even further. As LWRs use less than 1% of the energy in the natural uranium, there is substantially greater scope for improvement in the future. There are no physical impediments to extracting practically all the energy in the natural uranium by recycling the used fuel.

Through repeated recycling, the next generation of reactors (called fast reactors) will be able, in theory, to use all the natural uranium to generate heat. That would mean getting more than 100 times the energy from the natural uranium with a corresponding reduction in fuel used and waste created.

Fission energy has the unique advantage of using a fuel with an energy density millions of times greater than any other known energy source.[2]

14.4 Carbon Dioxide Emissions

Burning a tonne of coal releases around 2.5 tonnes of CO_2 so a 1,000 MW coal power station could discharge 7–10 million tonnes of CO_2 each year.

Using one tonne of uranium in a fission reactor produces no CO_2 at all so a fission power plant releases no CO_2 in operation. In a world where reducing greenhouse gas emissions is becoming ever more urgent that factor alone makes fission energy a very attractive alternative to coal energy.

14.5 Cost

Recently I, along with some colleagues, prepared an analysis of low-carbon baseload technologies by analyzing a wide range of baseload technology assessments that have been published over the last decade [1]. We included both nuclear power as well as clean coal and gas technologies using CCS. We also included solar thermal with molten salt storage and gas backup. Fission energy turned out to be the cheapest option and best able to meet the IPCC timetable for greenhouse gas reductions (see Chap. 6).

Fission was 30% cheaper than the least expensive CCS solution and only a third of the cost of the solar thermal solution.

14.6 What's the Drama?

Fission energy is one of the cheapest sources of electricity, it has a high capacity factor so it can provide baseload power, it produces no emissions in electricity production, and has a very low full life cycle emissions intensity [1]. So why do some in the community still resist using it? We will deal with that issue in Chap. 15.

Reference

1. Nicholson M, Biegler T, Brook B (2010) How carbon pricing changes the relative competitiveness of low-carbon baseload generating technologies. Energy. doi:10.1016/j.energy.2010.10.039

---

[2] There is a table of energy densities shown in Table E.1.

Chapter 15
Safety First

Those who are cautious about nuclear power often quote safety as one of their concerns. These concerns usually involve the safety of the reactors, the safe storage of the waste, and the risk that somehow the nuclear fuel used in the reactor will finish up in a nuclear weapon of some kind. In this chapter, we take an objective and open look at these safety issues. We start with the thorny issue of reactor safety and discuss the three main reactor accidents in the 50-year life of nuclear power plants; at Three Mile Island, Chernobyl and Fukushima. We then look at the volume of nuclear waste and issues around its storage and finish by discussing the nuclear proliferation risk.

15.1 Reactor Safety

Sorting out fact from fiction is often difficult. One objective way to assess the safety of a technology is to look at history. Fortunately with fission reactors we now have quite a bit of history to draw on. We have been using fission reactors to generate electricity for over 50 years. In that time, there have been just three significant accidents at commercial power stations, and only one created a serious radiation hazard to the public.

The first accident was in 1979 at Three Mile Island in the US, which experienced a partial core meltdown. The accident was caused by a technical malfunction followed by human error. Cooling-water pumps failed, preventing the steam generators from removing the heat from the reactor. An operator error prevented the backup system from operating and the reactor shutdown as it was supposed to under such circumstances. However, a pressure-relief valve designed to release increased pressure in the reactor failed to close properly once the pressure reduced. The open valve allowed radioactive cooling water to escape the reactor vessel causing the reactor core to overheat. The radioactive fuel stayed contained in the reactor vessel. Some trace radioactive gases were vented to relieve pressure on the containment dome, but it was relatively minor, exposing people in

M. Nicholson, *The Power Makers' Challenge*, Green Energy and Technology,
DOI: 10.1007/978-1-4471-2813-7_15, © Springer-Verlag London 2012

the immediate vicinity of the reactor to similar level of radiation they would get from a chest x-ray—frightening at the time, but not damaging to people's health.

Even though the damage was contained, it did not stop anti-nuclear activists from demanding the shutdown of all nuclear power plants, and the event helped to stop the progress of building new nuclear plants in the US. As a result of the accident, existing and future plants were made safer and operating protocols were improved, and over the last 30 years there has not been a similar incident.

The second accident was at Chernobyl in Russia in 1986 and was much more serious than at Three Mile Island. The Chernobyl accident dispersed part of the radioactive core into the environment as airborne dust and a cloud of fallout drifted over Europe. Human error again played a part, but in this case the reactor was flawed in design and of a type that was never built in the West. The accident happened as a result of what seems to have been a badly managed system test of some of the electrical systems, which involved deactivating safety systems. System warnings were ignored by the operators and when they eventually tried to control the problem it was all too late. The reactor went out of control in a way a light water reactor would never do under the same circumstances. A steam explosion (not a nuclear explosion) blew the top off the reactor and without a containment building (which is mandatory in most countries) radioactive material escaped into the environment for several days. Thirty-one people died at the reactor site and further 16 accident cleanup workers later died by the exposure to the extremely high levels of radiation in the immediate vicinity. It is possible that others outside the reactor compound developed cancer induced by the radiation and these could theoretically number several thousand, but this is difficult to confirm with any certainty. The official World Health Organisation death toll from the accident and its long-term aftermath remains at less than 100 individuals.

The third accident was at Fukushima in Japan in 2011. Japan was struck by a magnitude nine earthquake and 15 m tsunami. The Fukushima Daiichi nuclear plant, built 40 years ago, was positioned on the coast and met the full force of both the earthquake and the tsunami. The plant had been designed to withstand the equivalent of a magnitude five earthquake and 5–6 m tsunami. The earthquake triggered the boiling water reactors in the plant to shutdown the nuclear fission reactor automatically, as they were designed to do in an earthquake. The backup diesel power supply started automatically as planned and provided essential cooling water to remove decay heat from the shutdown reactors and spent fuel pools.

About an hour after the earthquake struck, the tsunami hit the plant. Because the tsunami was significantly higher than the protective wall, it swamped the plant and washed out the backup generators and fuel supply. The second line of defense was emergency battery power which operated for about 8 h but eventually all power was lost. The resulting total loss of power meant the loss of cooling water to the reactors and spent fuel pools causing them to overheat. It took several hours to replace the backup generators and eventually some of the fuel in the reactors melted, generating hydrogen. The venting of this hydrogen caused explosions in some of the reactor containment buildings resulting in significant building damage.

As with Three Mile Island, some radioactive gas was vented from the plant but outside the plant gate the radiation levels remained relatively minor with negligible health risk [1]. The venting of steam also carried away radioactive fission products, leading to a release approximately 10% of the release from Chernobyl. As a safety precaution, the residents living around the Fukushima plant were evacuated.

There are many lessons to be learned from the Fukushima accident. The plant was located in an area prone to earthquakes and tsunamis and inadequate attention was given to the location of the critical backup power supplies. The accident led to calls for other nuclear plant operators around the world to review their own protection against natural disasters.

Sad as industrial accident deaths are, in order to make a proper comparison with other energy technologies, we really have to look at the number of deaths from other energy sources, particularly coal, hydro, and gas. Thousands have died in coal mines and hydro dam accidents. This relatively low number of deaths from one nuclear accident after 12,700 reactor-years of civil operation attests to nuclear's relative safety and, apart from at Chernobyl, no plant operators or members of the public have ever died from exposure to radiation from a civil nuclear power plant.

A European Commission ExternE study [2] examined the external costs[1] of electricity generation using a life cycle assessment. The results show that, on a unit of energy produced basis, the health and safety costs of uranium mining and the use of nuclear fuel, including waste disposal, are significantly lower than for fossil-fuel-based energy generation.

Another way to assess reactor safety is to look at the probabilistic risk. Such an assessment was made in 1991 by the Nuclear Regulatory Commission (NRC) which found the risk of death or cancer from nuclear plants was extremely low (less than one in 10 million per year). This study is currently being updated.

The reactors being built today have passive[2] safety features which do not require an external power source or operator intervention in the event of an accident (see Appendix G—Nuclear Power). It seems likely that the accidents at Three Mile Island and Chernobyl were triggered by operator error. Passive safety features, which bypass the operators, might have prevented both these accidents. The Fukushima accident was caused by the loss of power that would not have the same impact on a modern passive reactor.

There is a public misconception that any level of radiation can be dangerous. This is clearly not the case. We are surrounded by natural radiation all the time, often at levels higher than allowed around nuclear power plants. Exposure to radiation from coal power plants can be over a 100 times higher than from nuclear

[1] External costs to climate, human heath, crops, structures, biodiversity, etc. that are not accounted for in the electricity cost.

[2] Passive means that no operator action is required to trigger the safety feature and they do not rely on an external power source to work.

plants [3]. Yet coal plants are not subjected to the same tight radiation regulations as nuclear plants.

All industrial processes are dangerous to some degree and fission energy is no exception. Based on the ExternE analysis, fission has lower external costs than the existing processes used to generate most of our electricity. It is safe to say that fission energy presents no more health risk to the public or the industry's workers than hundreds of other industrial and commercial activities that we take for granted.

15.2 Nuclear Waste

As we discussed in the Chap. 14 the volume of waste from fission plants is not really the issue. The concern is around the safety of transporting and storing that waste. The waste we are really concerned about is the long-lived (over 30 years) component of radioactive waste from spent fuel which needs to be isolated for a very long period of time. A 1,000 MW nuclear power station will produce about 30 t of high level solid packed waste per year [4]. This actually is not very much, under two cubic meters and will fit inside a telephone box. For comparison, a 1,000 MW coal power station will product 10,000 times that amount of ash waste in a year—about the volume of a 30 storey building.

Transporting radioactive material to and from power plants, as well as to and from medical and industrial sites, has been performed safely for decades. Only a small fraction of these consignments are related to nuclear power plants.

Millions of consignments of radioactive material are transported every year by road, rail, and ships. If an accident happened to one of these trucks, the danger from the diesel in the fuel tank exploding is much greater than the danger from leaking radiation. Of course, this applies to all truck transport irrespective of cargo. A nuclear explosion with this transported material is physically impossible.

The real issue is where to store the long-lived nuclear waste safely. Most of the waste produced in power plants over the last 50 years has been held close to the reactor site. Sometimes some of it is reprocessed but that does not significantly reduce the waste problem. Eventually, some of it will need to find a permanent and secure home. Current plans are to bury it deep underground in a safe repository. But, unlike the coal ash, not all (indeed, hardly any) of the energy has been removed from the spent fuel and this can be recovered in fast reactors. Uranium ore from the earth will not last forever. As currently mined and used it may not last out the rest of this century, so we may need this spent fuel to run the next generation of fission reactors. The "permanent" storage may not need to be that permanent—we should be able to recycle most of that spent fuel into electricity but that technology could be a couple of decades away from widespread commercial deployment (see Appendix G—Nuclear Power). We should actually see this 'waste' as a fuel store for our grandchildren and their grandchildren.

15.3 Proliferation of Nuclear Weapons

Nuclear proliferation generally refers to the spread of nuclear material, technology, or weapons to nations outside the five recognized nuclear weapon states of the US, Russia, the UK, France, and China. A few nations outside this group, India, Pakistan, Israel, and North Korea have developed weapons but not from civil nuclear power plants used to generate electricity.

Reactor fuel and waste from civil power plants is not a great place to start for a rogue nation or terrorist group wanting to make a nuclear weapon. Converting reactor-grade material to the weapons-grade needed for bomb making is difficult and expensive and would be readily detected. It would be cheaper and easier to buy or steal weapons-grade material from one of the above nations who already have it. And they have plenty of it, so it seems unlikely that access to a civil reactor would increase the risk of proliferation. Those nations hell bent on developing nuclear weapons will do so with or without fission energy. The cat may be out of the bag as far as nuclear proliferation is concerned, unfortunately.

A greater risk, although still not as risky as it may sound, is building a dirty bomb using radioactive material stolen from a nuclear power plant. A dirty bomb combines conventional explosive material (such as dynamite) with radioactive material, with the intension of dispersing radiation over a large area. This is obviously a concern but it will be the explosion that causes most of the damage not the radiation and fear will be the greater threat than the actual radiation exposure. Of course, such a device could also be produced using radioactive material from medical or industrial sources where they might be more readily available and more easily dispersed than the heavy metals from a fission power plant, so shutting down nuclear power plants is not going to eliminate the risk of dirty bombs.

The inconvenient truth is that banning nuclear power will neither reduce the risk of proliferation nor the risk of terrorism.

15.4 Safety Summary

Yes there are some dangers with fission energy, but there is with all industrial processes. The evidence suggests the dangers to health from fission energy are significantly less than the dangers from mining fossil fuels and breathing the pollution created during converting the fuel into energy—either in a power plant, or in a motor vehicle.

Even some renewable energy sources can be much more dangerous than nuclear power. The most common source of low-emission renewable energy, hydropower, has killed thousands of people through dam failures in many parts of the world. Accidents happen in all generating plants, including solar thermal plants and wind farms. Wind turbines can throw blades or parts of blades hundreds of meters and cause serious casualties and possible deaths. Wind farm construction and operation has already caused more deaths worldwide than nuclear plants even though wind power has produced a small fraction of the energy produced by nuclear power [5].

If we walk away from fission energy, as advocated by some environmentalists, we will continue to burn fossil fuels, which the evidence says is not only more dangerous to our health and safety but a much greater threat to climate change.

References

1. MIT Advanced Nuclear Energy Systems (2011) Technical lessons learned from the Fukushima-Daichii accident and possible corrective actions for the nuclear industry: an initial evaluation. http://mitnse.files.wordpress.com/2011/06/fukushima-lessons-learned-mit-nsp-025.pdf. Accessed 24 July 2011
2. European Commission—ExternE (2003) External costs: research results on socio-environmental damages due to electricity and transport. http://externe.info. Accessed 24 July 2011
3. Gabbard A (2008) Coal combustion: nuclear resource or danger. http://ornl.gov/info/ornlreview/rev26-34/text/colmain.html. Accessed 24 July 2011
4. IAEA (2011) Factsheets, managing radioactive waste. http://iaea.org/Publications/Factsheets/English/manradwa.html. Accessed 24 July 2011
5. Rockwell T (2010) Nuclear energy facts report. http://learningaboutenergy.com/nuclear-energy-facts-report/. Accessed 24 July 2011

Chapter 16
More Smoke and Mirrors

In Chap. 11, we looked at some comparison myths around renewable energy. In this chapter we look at some myths that seem to surround fission energy. For some reason nuclear power attracts a great deal of flack. As discussed in the Preface, it's probably because of the fearful association with nuclear weapons and the damage they have done to innocent people. However, if we are to tackle climate change in a realistic way, we need to get over these irrational fears. Tackling climate change means moving away from fossil fuels over the next few decades. In most countries, fission energy will be the only way they can do this without impacting the reliability of their electricity network. In this chapter we cover several issues including, need, expense, time to build, sustainability, radiation, and investment risk.

16.1 Myth 1: Renewables Make Nuclear Unnecessary

In many ways this is one of the most dangerous myths because it can seem so seductive. We have extensively covered this in Part II—Renewable energy by systematically reviewing all the renewable options and seeing if any of them are capable of replacing coal (or fission), which currently supplies the bulk of the electricity in most countries. The outcome was that no scalable renewable source is capable of replacing coal throughout much of the world without some technology breakthrough—probably in energy storage.

The reason this is a dangerous myth is it puts false hope in the minds of people and politicians that, somehow one day we can replace all that dirty coal with nice clean renewable energy. Politicians, in good faith, are spending tax payers' money on financial incentives to RE developers to build wind and solar farms that so far have not shutdown a single coal plant and probably never will.

Investing in fission energy would not only save money (as it is cheaper than wind and solar) but it would immediately eliminate the need to build new fossil-fuel plants and allow the old ones to be progressively closed. This will have a much greater impact on climate change than building wind and solar plants.

M. Nicholson, *The Power Makers' Challenge*, Green Energy and Technology,
DOI: 10.1007/978-1-4471-2813-7_16, © Springer-Verlag London 2012

16.2 Myth 2: Nuclear Energy is Too Expensive

Fission plants are expensive to build but cheap to operate compared to fossil-fuel plants. This means the cost of the borrowed money (the interest charged) to build the plant has a greater impact on the cost of electricity than for fossil fuel which is more heavily affected by the cost of fuel.

Once a fission plant has been operating for a number of years, and the initial construction costs have been fully recovered, the operating cost of the plant per unit of energy produced is lower than for fossil-fuel plants and many RE plants.

As discussed in Chap. 5, the cost comparison between energy sources has to be based on the full life cycle cost and total energy produced. Just because fission plants require a significant upfront capital investment does not necessarily make them expensive over the plant's lifetime. In Chap. 14 we found that nuclear is the cheapest low-carbon baseload energy source once an adequate carbon price is applied.

What does need to be avoided is charging higher borrowing costs for fission plants than for other plants, based on spurious assumed risks. This artificially increases the cost of fission energy and only makes more profit for the plant financiers. The perceived risks of nuclear power has led some countries like the US to offer government loan guarantees to private financiers to encourage them to invest in new fission plants.

It is often claimed that nuclear energy gets more government financial support than renewable energy. This may be true in total dollars, but the real measure is dollars per unit of energy produced by the supported technologies. In the US at least, the dollars per MWh allocated for wind and solar energy is over 20 times the dollars allocated to fission energy [1].

16.3 Myth 3: Nuclear Plants Could Not be Built in Time

It is true that fission plants take a number of years to build. Some of this time is in seeking and receiving government approval which could be fast-tracked—as happened in France. But building large plants does take time. We need to urgently move to low-carbon electricity sources so the argument goes that building new nuclear plants will take too long.

But what is the alternative? We have learned by now that wind power and biomass cannot replace the coal plants and no one seems to want more hydro plants. Even if they did the hydro plants also take a long time to build. We could replace the coal plants with gas turbines. In some countries, like Australia, this is taken seriously as the most attractive option. Australia is blessed with abundant supplies of gas, but it will not last forever and will get more expensive over time. Furthermore, gas will not achieve sufficient emission reduction as we discussed in Chap. 13.

The only viable low-carbon, non-fission alternative that might be scalable to replace coal plants is solar thermal with heat storage, located in the solar belt. But these plants will also need to be huge. They require eight times as much concrete and 13 times as much steel as nuclear plants for each GWh generated (see Table 11.2). If we cannot build the fission plants in time then we surely cannot build the solar thermal plants in time either.

16.4 Myth 4: Uranium Supplies are Not Sustainable

By definition, the energy source for renewable energy (primarily the sun) is sustainable, whereas anything mined from the Earth will run out eventually. This is very clearly the case for coal, oil, and gas. But uranium and some other fission energy fuels like thorium are very different to fossil fuels. As discussed in Chap. 14, current thermal reactor technology uses less than 1% of the uranium's energy potential.

Certainly today, where most uranium is used in light-water reactors without recycling, the supply is not sustainable. Current identified high-grade resources will last about 85 years. Total expected resources will last longer at 270 years—but this is still not sustainable long-term [2].

Uranium is so cheap today that we do not bother to try and recover the other 99%. It becomes part of the infamous nuclear "waste". With reprocessing and fast reactors we can release nearly all the energy from the mined uranium or thorium which will extend the known mined reserves to tens of thousands of years. Fast reactors will also produce fuel for the existing fleet of thermal reactors. And ultimately, should all other sources run out, there is plenty of uranium dissolved in the oceans that could, technically, be extracted and used for fuel.

16.5 Myth 5: All Radiation is Dangerous

Radiation can be dangerous in high doses. That is why medical procedures involving radioactive materials are conducted with great care. The same is true in a fission energy plant. High safety standards and procedures ensure that the public and the plant operators are not exposed to dangerous levels of radiation. Modern plants have multiple layers of safety features to protect against operator error and to make sure there are no radiation leaks. The safety history of nuclear power plants over the last 50 years is evidence of this.

Some would argue that radiation and radioactivity are unnatural. But soil, water, the air we breathe, and even our own bodies are all naturally radioactive. Far from being dangerous, it is possible that radioactivity in low doses may actually be beneficial [3].

Fission changes long-lived radioactive uranium into shorter lived fission products so actually reduces the Earth's radioactivity. Even if we were adding to the radioactivity it would be more than offset by the Earth's natural radiation decay. Every year the Earth becomes a little less radioactive, not more so.

16.6 Myth 6: Nuclear GHG Emissions are Huge

This myth usually refers to the GHG emissions to build the nuclear plant and mine the uranium. Fission energy produces no emissions in operation. As with comparing costs, it is the life cycle emissions (emission intensity) that are important, not the emissions released to provide the raw materials and to build the plant.

Table 6.1 shows that nuclear has an emission intensity as low as any renewable energy source.

16.7 Myth 7: Gas is Less Risky for Investors

At first glance this might seem reasonable. Gas plants are a mature and reliable technology with a lower emission intensity than coal (but not as low as nuclear). Gas does not have the regulation requirements and higher borrowing costs of nuclear despite being inherently more dangerous (gas is very explosive as was demonstrated in a recent new Connecticut US gas plant accident which killed at least five people [4]).

The drawback with gas is the uncertain future fuel price which will increase electricity generating costs and the introduction of a carbon price which will also significantly increase generating costs. Fission energy has neither of these problems, and will actually be a more certain investment than any fossil-fuel plant once a price is put on carbon.

References

1. US DOE/EIA (2007) Federal financial interventions and subsidies in energy markets Table 35. http://eia.doe.gov/oiaf/servicerpt/subsidy2/pdf/chap5.pdf. Accessed 25 July 2011
2. World Energy Council (2007) Survey of uranium resources. http://worldenergy.org/publications/survey_of_energy_resources_2007/uranium/673.asp. Accessed 25 July 2011
3. Rockwell T (2010) Nuclear energy facts report. http://learningaboutenergy.com/nuclear-energy-facts-report/. Accessed 24 July 2011
4. Reuters (2010) At least 5 dead in Connecticut gas plant blast. http://reuters.com/article/2010/02/07/us-energy-explosion-idUSTRE61619Q20100207. Accessed 25 July 2011

Chapter 17
Is Fission Really Necessary?

If you have been reading this book right through so far you have probably already worked out the answer to the question posed in this chapter heading. If you are concerned that man-made CO_2 is having a significant impact on climate change and you want to keep those changes to an absolute minimum, then effective action to replace fossil fuels for electricity generation needs prompt attention. If we want to maintain a reliable and inexpensive electricity supply and replace fossil fuels then, like it or not, we must use fission energy. This chapter revisits the evidence so far that supports this bold statement and discusses how the Power Makers will react to the challenge of emissions reduction.

17.1 Challenging Times

In Chap. 6, we discussed climate change and what the IPCC scientists say about the need for greenhouse gas emission reductions. In summary, we must reduce our emissions by 85% by 2050 to have a reasonably high chance of avoiding major climate change. Electricity generation is a major contributor to greenhouse gas emissions, so we must target to reduce emissions from electricity generation by 85% at least by 2050.

To reach the emissions reductions needed by 2050 the worldwide average emission intensity of electricity generation needs to be reduced to 50 kg CO_2-e/MWh. Currently it stands at about 10 times that figure.

Reducing emission intensity demands that the Power Makers change the fuels they use to generate electricity. This means replacing plant that may still have a long useful life so they will not do that willingly. The Power Makers will probably be pushed into it by the introduction of a carbon price which will penalise them for every tonne of CO_2 they emit and so force them to raise their electricity prices. This carbon price will need to progressively increase as emission intensity targets are tightened.

M. Nicholson, *The Power Makers' Challenge*, Green Energy and Technology,
DOI: 10.1007/978-1-4471-2813-7_17, © Springer-Verlag London 2012

17.2 Renewable Energy

The Power Makers will be forced to reduce emissions but they will still be required to deliver a reliable electricity service. They will only be looking at technologies that can reliably replace their baseload coal plants which are the plants that generate the most CO_2.

The Power Makers are probably well aware of the limitation of renewable energy we discussed in Part II—Renewable Energy. Difficult-to-dispatch renewable sources like wind and solar PV will not replace much baseload coal generation. Where they have access to dispatchable renewable resources like geothermal, biomass and hydro, they will use them to the fullest extent possible as they become cheaper than coal including a high enough carbon price. In some networks it may even be possible to just use dispatchable renewables, but these will be in the minority. Most Power Makers will need to look elsewhere for technologies to replace coal.

17.3 Clean Coal

For some Power Makers, carbon capture and storage could be the answer, but it might be another decade at least before this technology is ready to be used on the commercial scale needed. This means delaying the change and possibly having to pay a hefty carbon price in the meantime.

Some will elect to not wait for CCS but switch from coal to gas instead. The Power Makers know this is not a long-term solution and will require a significant investment in new gas plants. Even if they switch to gas, they will still be faced with the need for CCS as the carbon price continues to rise.

New coal and gas plants can be expected to have a 40-year life. CCS will not capture all the emissions so the Power Makers will still have to pay some price to emit the albeit substantially reduced CO_2 that the CCS plants will release. All the CCS options whether using coal or gas have a higher emission intensity than the 2050 target (see Table 6.1) so relying on CCS will make that target difficult to achieve.

17.4 Fission Energy

Fission plants can handle baseload as well as any coal or gas plant so reliable supply is not a concern. Fission plants produce no CO_2 in operation so a rising carbon price is not a concern either. They can be built near the site of the existing fossil-fuel plants so there are no expensive long-distance transmission upgrades as demanded by most renewable energy plants.

Many Power Makers will elect to use this option unless stopped by community concern. Any other option will either add to the cost of electricity or compromise reliability in the network or both. China is leading the fission energy pack today with almost half the new reactors (25 as of 2012) being built there.

Communities and their elected political leaders will need to decide which option delivers the best long-term outcome. Every political decision comes at a cost of some sort. Perhaps the NIMBY[1] and global anti-nuclear activist groups will reconsider there position once the true cost of the non-nuclear alternatives is recognised.

[1] Not in my back yard.

Chapter 18
The Next Generation of Fission

Like most energy sources, fission energy is constantly evolving but fission is the one energy source that has the capability of supplying our electricity demand indefinitely at lowest cost without producing greenhouse gas emissions. In this chapter we look at some current developments and see how we can make fission energy into a truly sustainable energy source. We cover the issues of fuel recycling, newer reactor technology, modular, and small reactors and finish with a discussion of fast breeder reactors to make fission energy sustainable.

18.1 Fuel Recycling

Current generation thermal reactors with no fuel recycling use less than 1% of the energy potential of natural uranium. Because uranium has been plentiful and relatively inexpensive, most nuclear waste from thermal reactors is not recycled. At current uranium prices, attempting to extract more energy from the uranium by reprocessing the waste (see Sect. G.3 in Appendix G—Nuclear Power) has not been considered cost effective in many parts of the world.

As the world's stocks of high-grade uranium are used up and become more expensive there will be more interest in reprocessing thermal reactor fuel to extract 25% more energy from the uranium [1].

Improving the energy efficiency by 25% sounds good but it does not turn uranium into a long-term sustainable energy source. We need fast reactors and improved recycling methods for that and we will discuss these toward the end of this chapter.

18.2 Generation III Reactors

Most currently operating fission reactors are considered to be Generation II reactors like the light-water reactors we discussed in Chap. 14. The new generation of reactors being installed today are called Generation III reactors. These comprise

M. Nicholson, *The Power Makers' Challenge*, Green Energy and Technology,
DOI: 10.1007/978-1-4471-2813-7_18, © Springer-Verlag London 2012

several advanced light-water reactor designs that provide standardized designs, better economics, and passive (rather than just active) safety features that do not require an external power source or operator intervention in the event of an accident. This makes fission reactors even safer. They also have better fuel utilization and slightly reduced waste.

The Westinghouse AP1000[1] reactor is claimed to be the safest and most economical nuclear power plant available today. The plant design is simplified to speedup the construction time and to make it easier and less expensive to build, operate, and maintain. It also includes passive safety features.

Future generation thermal reactors may use gas (probably helium) rather than water as the working fluid. The helium would drive a gas turbine rather than generate steam for a steam turbine. The fuel for these reactors can be made much more proliferation resistant. They run at much higher operating temperatures that convert more of the fuel into energy and produce less long-lived waste (see Sect. G.4).

18.3 Modular Construction

Historically, fission reactors have been constructed on-site, often using purpose built designs. It could take over 5 years to build a new reactor. The high cost of building reactors has been in the construction labor, not in the building materials, which are much less per unit of plant energy generated than many other technologies (see Table 11.2).

There is a move to make modern reactors modular using standardized designs. This means they can be constructed more efficiently in factories as smaller modules that can be shipped and assembled on-site. This reduces the construction time down from over 5 to 3 or 4 years. It also reduces the amount of construction labor and reduces the investment period, lowering costs.

Most large fission power reactors produce about 1,000 MW or more. There has been interest in building small modular reactors (SMR) that could be from 10 to 300 MW. These compact designs could be made in factories and transported to sites as an entire unit, by truck or rail. They could be connected at a single site to scale-up to a larger power station or used for distributed generation. A type of thermal SMR is already in use in nuclear submarines.

18.4 Fast Reactors

The next generation of reactors (Generation IV) will include fast reactors. Fast reactors still produce energy using fission but they do it in a way that allows them to convert a much higher proportion of the uranium into energy (see Sect. G.6).

[1] AP stands for "advanced passive", and 1,000 is the MW power output.

Fast reactors use high energy fast neutrons (hence the name "fast" reactor) in the fission process rather than the slowed-down neutrons used in thermal reactors. This allows them to extract energy from all the natural uranium not just the small percentage available to thermal reactors. Potentially they can extract 160 times the energy from uranium than is possible with open cycle (no recycling) thermal reactors. In practice it may be less than half of this but it will still be substantially more than is possible with thermal reactors using fuel recycling. Fuel recycling is an integral part of fast reactors.

Fast reactors can also extract energy from the waste from thermal reactors as well as produce new fuel for thermal reactors. So the past 50 years of accumulated uranium and plutonium waste ("spent fuel") could be turned into energy rather than buried in the ground. A fast reactor can extract from a kilogram of uranium waste the same energy as contained in 11,000 barrels of oil. At $90 a barrel, each kilogram of waste is worth nearly a million dollars in energy value! The nuclear waste stockpile is growing at around 12,000 t a year—that is over a trillion dollars in energy value. It is estimated that nuclear fission fuel is, for all practical purposes, "inexhaustible" if used in fast reactors [2]. Inexhaustible is certainly sustainable.

Fast reactors have so far not been commercially competitive with thermal reactors and have not been widely used for electricity generation. There is still some debate about just how long it will be before fast reactors become commercially viable. This may depend, among other things, on the availability and relatively low cost of uranium for use in thermal reactors. As uranium supplies start to diminish toward the end of the century, we can expect to see a greater interest in fast reactors. France, Japan, India, and Russia all have fast reactors in operation for research [3], and South Korea is seriously considering recycling its spent fuel stockpiles in fast reactors rather than depositing it in permanent storage. France has plans to replace all its thermal reactors with fast reactors before the end of the century and China is also planning to use fast reactors. The notable exception is the US, which closed its last fast reactor in 1994.

There are even plans to build a fast SMR that is sealed with no moving parts and small enough to be portable and trucked to site and buried underground. The module would contain enough fuel for 7–10 years and be returned to the manufacturer for refueling. Each module would be 25 MW so is large enough to run a factory or a medium-sized town of 20,000 homes [4]. Other larger fast SMRs are also under developed [5].

References

1. World Nuclear Association (2011) Processing of used nuclear fuel. http://world-nuclear.org/info/default.aspx?id=466&terms=reprocessing. Accessed 25 July 2011
2. Lightfoot D, Manheimer W, Meneley D, Pendergast D, Stanford G (2006) Nuclear fission fuel is inexhaustible, IEEE Xplore. http://computare.org/Support%20documents/Fora%20Input/CCC2006/Nuclear%20Paper%2006_05.htm. Accessed 25 July 2011

3. World Nuclear Association (2011) Fast neutron reactors. http://world-nuclear.org/info/default.aspx?id=540. Accessed 25 July 2011
4. Hyperion Power (2011) A new paradigm for power generation. http://hyperionpowergeneration.com/product.html. Accessed 25 July 2011
5. World Nuclear Association (2011) Small nuclear power reactors. http://world-nuclear.org/info/inf33.html. Accessed 25 July 2011

Part V
The Melting Pot

Energy access for all will require making available basic and affordable energy services using a range of energy resources and innovative conversion technologies while minimizing GHG emissions

Intergovernmental Panel on Climate Change (2007)
WGIII Ch 4 Executive Summary

So far we have talked about three groups of electricity plants: fossil fuel, fission, and renewables. Historically, most countries have largely relied on fossil-fuel plants with some renewables like hydro, geothermal, and biomass. Over the past few decades fission has taken over from some of the fossil-fuel plants. There has also been a steady increase in the number of wind and solar plants although they still represent less than 3% of the total electricity production.[1]

However, as the pressure increases to reduce greenhouse gas emissions from electricity generation, there will be an accelerated push to replace baseload coal plants with fission plants as the most cost effective low-emission baseload generators. But fission cannot run the whole network. As we learned in Chap. 4, electricity supply must match the load moment by moment and the Power Makers need flexible generators to follow this load as it rises and falls. They also need generators that can be started quickly when needed (standing reserve). Fission plants can do some of this but, like coal, they take some time to get going from a cold start so cannot be started quickly if suddenly needed to meet a jump in demand.

We can expect to see, in the future, a mix of generating technologies much as we do today. The Power Makers prefer not to be too dependent on any one fuel source, be it coal, gas, oil, uranium, hydro, solar, or wind. The actual mix will change as the pressure to reduce greenhouse gas emissions drives up the carbon price, forcing the Power Makers away from coal and gas. It will also change as we consider alternative network strategies such as decentralization, smart grids, and more island networks. We will discuss all these in the final part of this book.

---

[1] IEA (2010) Key World Energy Statistics 2010 http://iea.org/publications/free_new_desc.asp?pubs_ID=1199 Accessed 10 August 2011

Chapter 19
Variable Loads are Here to Stay

We start this final part of this book by looking at variable loads. Even if all our baseload demand came from fission energy we still need to service the demand above baseload—referred to as intermediate and peak load in Fig. 19.1. In this short introductory chapter we revisit intermediate and peak loads and how they are serviced. We then discuss the on-going need to address variable loads into the future.

19.1 Intermediate and Peak Loads

Let us recap what we learnt in Chap. 4. Electricity demand is generally categorised as baseload, intermediate load and peak load. Baseload takes up most of the load and does not tend to change significantly over time. It is really the minimum amount of power the Power Makers must always make available to its customers. Intermediate load does vary but is predictable and influenced by time of day such as weekday mornings and evenings. Peak load is much less certain and is often influenced by climatic conditions, such as cold snaps or heat waves, that change demand for building heating and cooling.

The Power Makers use different generators to service the three different loads. Typically coal, gas and nuclear are most commonly used for baseload, but some lucky countries have plenty of hydro power (e.g. Canada, Brazil, Norway) which they can use instead. Conventional geothermal and biomass are also used for baseload in some countries (e.g. Iceland).

Intermediate load is relatively predictable and occurs around 06:00 (6 a.m.) and continues into the evening. It can be serviced by gas-heated steam generators, gas turbines, hydro plants as well as solar thermal plants with heat storage or gas hybrid systems (see Appendix D—solar power). Many coal and fission plants also service intermediate load. These generators need to be able to ramp up and down to follow the load as it increases and falls through the day. For further discussion on how this is done see Sect. B.3 in Appendix B—Power Generation and Network Control.

M. Nicholson, *The Power Makers' Challenge*, Green Energy and Technology, DOI: 10.1007/978-1-4471-2813-7_19, © Springer-Verlag London 2012

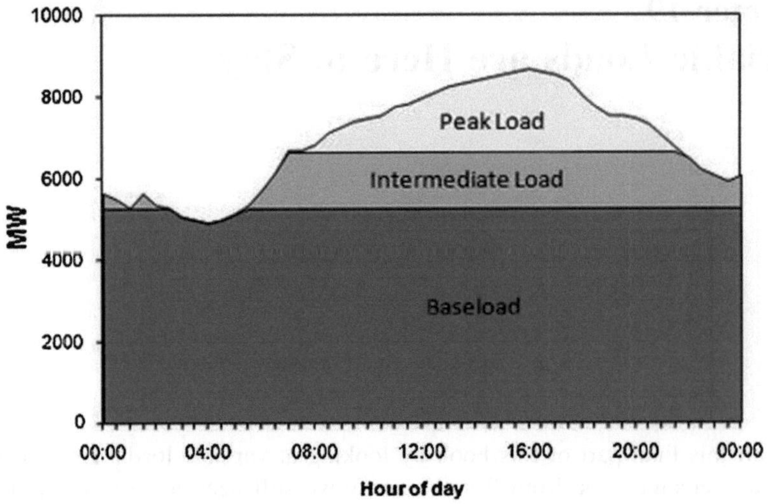

Fig. 19.1 Electricity load curve

The most variable part of the total demand is peak load. The Power Makers attempt to predict what this might look like 24 h ahead based on weather forecasts and expected, but infrequent, events (such as a large public event) so they can schedule generators to be available when needed. They still need some unallocated standing reserve that can be connected and ramped up quickly to handle unexpected increases in peak demand and they may need to request that some scheduled generators do not run because the anticipated peak failed to appear. All this still requires predictable generators that will be there when you need them. This makes variable resources like wind and solar PV difficult to schedule for this task, but with improved forecasting this may become easier.

19.2 Variability Won't Go Away

Some hope that with more demand management, where the system takes greater control over when power is used, it may be possible to smooth out the load and reduce the peaks. This might ease the Power Makers' job of servicing the variable load. We briefly discussed demand management in Chap. 10.

Some future developments like the electrification of transport may help with smoothing the load. Domestic electric vehicles are likely to be charged when the owner is sleeping and the electricity demand to recharge vehicle batteries could be a significant part of a household's total electricity usage. This will tend to even out the difference between daytime and nighttime load.

There are clearly electricity demands that can be time shifted, but it is hard to imagine a time when we will use the same amount of electricity during the night as we do during the day. Humans are likely to continue to be more active during the day than at night and hence use more electricity in daytime. Demand response for managing peak loads involves, at most, 100 or so hours a year (about 1%), and seldom more than a few consecutive hours [1]. But any reduction in peak load from demand management will reduce (but not eliminate) the need for reserve generator capacity. It will also reduce the peak load demand on the transmission and distribution network.

We will continue to need generators that can reliably deliver much of our electricity at low cost (baseload) while using other generators to follow the rising and falling load during the day—again at the lowest cost. We discuss the impact of generating cost on generator usage in Chap. 20.

Reference

1. US DOE (2010) The smart grid: an estimation of the energy and CO_2 benefits. http://energyenvironment.pnl.gov/news/pdf/PNNL-19112_Revision_1_Final.pdf. Accessed 25 July 2011

Chapter 20
Cost Matters

Power Makers are expected to generate sufficient electricity at the lowest possible cost. This means generator costs play a key role in deciding which generators are used to meet the loads. In this chapter we compare generator costs in the US. We then consider the impact that these costs have on the generator mix used by the Power Makers as well as the impact of a rising carbon price.

20.1 Comparing Generator Costs

Over the life of a power plant, what matters to the generator owners is the levelized cost of electricity (LCOE). We briefly discussed LCOE in Chap. 5 and describe it in some detail in Appendix H.

Table 20.1 shows cost data published by the International Energy Agency [1] for various plants built or being built in the US. These costs include a carbon price of $30 per tonne of CO_2-e. Long term, this carbon price will be on the low side but it helps to more realistically compare the costs of low and high CO_2 emitting technologies. These costs do not include transmission costs or the system cost of providing additional reserves to backup RE sources like wind and solar PV. This means that the costs for these variable RE technologies may be understated. For CCS technologies, the cost of carbon capture and compression is included but not the cost of storage. The IEA anticipated that capture and compression will account for a large proportion of total CCS costs.

The table shows both a low and high cost. The low cost is based on a LCOE discount rate of 5% and the high cost on a discount rate of 10%. See Appendix H—Electricity costs and Markets for a discussion on LCOE discount rates.

Remember these costs are for specific plants in the US and the IEA 2010 report makes it clear that real costs vary significantly between countries so we can expect that different countries will use different technologies, depending on their costs and availability of resources.

M. Nicholson, *The Power Makers' Challenge*, Green Energy and Technology,
DOI: 10.1007/978-1-4471-2813-7_20, © Springer-Verlag London 2012

Table 20.1 Levelized cost for new US plants in 2010

Energy source	Low cost US $ per MWh	High cost US $ per MWh
Coal	72	83
Coal—IGCC	75	92
Coal—IGCC with CC(S)	68	93
Gas—CCGT	77	83
Gas—CCGT with CC(S)	92	104
Nuclear—Adv Gen III+	49	77
Wind—onshore	48	70
Wind—offshore	101	146
Biomass	54	80
Geothermal—conventional	31	47
Solar PV	215	333
Solar thermal	211	323

From IEA [1]

20.2 The Generator Mix

Using the costs shown in Table 20.1, which include a carbon price of $30, it would seem to be financially attractive for the Power Makers to use on-shore wind power whenever it was available. They may, of course, be forced to revert to more expensive gas generators if the wind failed to deliver the energy when needed. They have to pay a price to have additional gas generators on standby to backup the wind. This additional cost is not reflected in the table and is not easy to assess because it very much depends on the location of the wind farm and the quality and quantity of the wind.[1] Wind power would probably not replace nuclear at the costs shown in the table, but some wind could well replace coal or gas in some locations, particular as the carbon price increases.

Power Makers use market mechanisms to determine which generators get used when. Most electricity networks have a trading market where generators can offer to supply a certain quantity of electricity in a given period during the day. The price offered into the market is based on the marginal cost of the generator. Marginal cost is the cost of producing an additional unit of electricity (e.g. MWh) and mainly reflects the fuel and additional operational costs.

Which generators are used in any given period is based on the price offered by the generator, with the cheapest getting first priority. This ensures the Power Makers are delivering electricity to the market's customers at the lowest possible cost.

The price actually paid to the generator operators (and charged to the wholesale customer) for electricity in that period is based on the highest-price generator used in that period. This gives incentive to the generator operators to have the most efficient equipment, as the lowest-cost generator earns the most profit. Note these

[1] It also depends on the total amount of energy that the wind is supplying to the grid (penetration).

dispatch periods can be quite short—from 5 to 30 min. Period prices can vary considerably throughout the day, with peak period prices many times greater than baseload prices.

For those who would like to know more about electricity markets, you can read Appendix H—Electricity Costs and Markets.

20.3 The Impact of a Rising Carbon Price

Table 20.1 above shows electricity costs based on a carbon price of $30. By 2040, it is quite possible that the carbon price will have risen to three times that figure. This will significantly increase the marginal cost of coal and CCGT plants (by roughly $90 and $40 per MWh, respectively). Power Makers will have even greater incentive to use fission and renewable energy plants as the marginal cost of these plants is only weakly impacted by the carbon price.

The rising carbon price will see a progressive shift away from using fossil-fuel plants and encourage their replacement by fission and renewable energy.

Reference

1. IEA (2010) Projected costs of generating electricity 2010 edition. https://iea.org/publications/free_new_Desc.asp?PUBS_ID=2207. Accessed 25 July 2011

Chapter 21
Distributed Generation and Storage

Distributed generation and decentralized energy are seen by some to be the answer to replacing large centralized generators. In this chapter, we look at the advantages and disadvantages of these concepts. We then proceed by discussing community energy storage systems and demand management to reduce local variability. We finish this chapter by considering the future of distributed generation.

21.1 Distributed Generation

The concept of Distributed Generation (DG) involves using a number of smaller generators located close to the end users. These generators might be stand-alone or integrated into a transmission network. These smaller generators can be built relatively quickly and can reduce the demand on the centralized generators.

There are some drawbacks to traditional "centralized" generation systems. For some countries it means they largely have to rely on imported fossil fuels, which produce environmental pollution and greenhouse gases and present energy security risks. They are often located some distance from the major load centers, so require long transmission lines with the associated transmission losses (see Chap. 3). Installing a number of small wind, solar, gas, diesel, biomass, or hydro systems or even small modular fission reactors (SMR—see Chap. 18) could serve local communities and reduce the reliance on large, fossil-fuel burning centralized generators.

In theory, if we had sufficient decentralized generators we would not need a big transmission network at all. Each "island" of locally connected generators could just support the local community. Just like they did a century ago. This is a very attractive option for those who like the concept of self-sufficiency. Some remote towns are still not connected to electricity grids and have to rely on local generators.

There are some drawbacks to DG. It could mean more expensive electricity to the end user and we would need more spare capacity in each island network to ensure reliability.

M. Nicholson, *The Power Makers' Challenge*, Green Energy and Technology,
DOI: 10.1007/978-1-4471-2813-7_21, © Springer-Verlag London 2012

Servicing just a local network with a few thousand users means that the electricity demand on that network will be much more varied (up and down) than for a much larger network with millions of users. Assuming we did keep some central generators, it would make the Power Makers job of controlling those DG networks much more difficult, requiring the use of much "smarter" networks. We will discuss smart grids in Chap. 22.

Some see DG as an early action response to reducing greenhouse gases while we transition the central generators from coal to cleaner sources like fission energy and renewable energy. This is probably a more realistic picture than looking to eliminate central power stations completely.

21.2 Decentralized Energy

Decentralized Energy (DE) is energy generated at or near the point of use. Decentralized energy is one aspect of distributed generation. The concept is that small (5 kW to 30 MW) distributed generators are connected to a local (island) distribution network, supplying homes and offices, rather than relying on the high voltage transmission network. These are sometimes referred to as mini-grids.

One benefit of decentralization is that smaller combustion power plants using gas, oil, or biomass can be used to provide both electricity and heating [often called cogeneration, trigeneration, or combined heat and power (CSP)]. These plants can operate more efficiently by not wasting heat as happens in large centralized power plants but directing that heat into buildings were it can be used for both heating and cooling. This is possible because decentralized generators are more likely to be close to buildings where the heat can be used, unlike large centralized power plants. Also, because the decentralized electricity is produced locally, the transmission losses can be kept to a minimum. That is the good news.

The not so good news is that despite the improved efficiency, small generating units typically cost [on an levelised cost of electricity see Chap. 20 basis] more per unit of electricity generated than large central generators—even with transmission losses. Some of this increase in electricity cost could be offset by savings in heating and cooling costs. The increased load variability across a smaller number of users requires more reserve capacity, demand management, and/or energy storage to ensure reliability of supply. Decentralized generation may also require significant upgrades to the local distribution network. This could all add to the cost.

DE island networks will still probably need connection to a centralized network to provide the reserve capacity and provide a market for surplus electricity supply. In this mode, they will not fully replace large centralized fossil-fuel or fission plants as these will still be needed for heavy industry and providing balancing reserves. If the DE island had its own energy storage, it might not need to call on the grid to provide external services.

21.3 Community Energy Storage

Energy storage can help with managing variability in load and supply. In a DE network, that storage would be attached to the local network, probably alongside a transformer or network substation. These systems could be relatively small, with clusters of batteries providing the 10–20 MWh of storage typically needed for a DE network. Smaller neighborhood storage systems of under 100 kWh could be used to service a small street community.

Plug-in battery electric vehicle (BEV) car batteries could also provide this storage service. A typical BEV has 25 kWh of storage in fully charged batteries. If these batteries were made available for community energy storage when the vehicles were parked in the garage, an average of four available BEVs could provide sufficient storage to manage a street community. Of course these vehicle owners would want to be confident that the vehicle batteries were fully charged when they needed to use the vehicle so the vehicle charging system would need to know the owners requirements.

21.4 Reducing Demand Variability

DE networks can benefit from using demand management to reduce load variability. Some loads, like refrigerators, can run at anytime so they can be automatically turned on or off depending on supply availability and so help with managing variability.

Demand management can also reduce peak demand on the DE network by shifting loads from high demand periods to low demand periods. This reduces the need for additional generator capacity during peak times. We discussed demand management in Chap. 10.

Both community storage and demand management require increased intelligence in the local network to ensure that storage systems are properly managed, but let us leave that until Chap. 22.

21.5 Distributed Generation Future

DG currently provides about 10% of the world's electricity generation but it is still early days. Some industries have been using local generators for many years. Cogeneration and trigeneration systems have also attracted some interest particularly in cold or heavily urbanized countries with densely populated cities. Hospitals and many large buildings have local backup diesel generators to provide emergency power supply during periods of grid supply interruption.

Building DG systems can involve big investment in new infrastructure—generators, storage systems, demand management systems, and upgraded distribution networks. DG is unlikely to be more widely adopted until it is cost competitive with existing central grid supply. Some governments will provide financial incentives or even legislate to encourage the adoption of DG. There may be some reduction in DG costs over time but, like renewable energy generally, it will stand or fall on being commercially viable over the long term.

Chapter 22
Smart Grids

The "smart grid" is often raised as a panacea for the use of variable renewable energy sources in the electricity network. In this chapter we discuss the smart grid vision along with the benefits and possible risks. Current electricity transmission systems use what is called a SCADA[1] system to monitor and control the transmission network from the generator to the distribution network substation. The system detects the need for any increase or reduction in generating resources and responds to system instabilities. These SCADA systems are relatively slow (they can take several seconds to respond) and have limited or no visibility in the distribution network beyond the substation.

22.1 The Vision

The smart grid aims to deliver electricity from suppliers to consumers using digital technology to optimize energy savings, energy cost, and reliability. It will use information technology to provide more visibility and control of both the existing grid infrastructure (both transmission and distribution) as well as customer demand response and distributed energy resources. The smart grid could extend all the way from the generator right down to the customer meter and even end user devices like air conditioners, pool pumps, clothes driers, and refrigerators.

Currently, the smart grid is really just a vision for a distributed, internet-like system that will provide better control of existing grid infrastructure, both transmission and distribution. It would integrate new as well as existing distributed assets (such as small local generators or storage systems) into the existing control systems (see image below). This requires upgrading and extending the existing SCADA system beyond the substation. The aim would be to accommodate all generation and storage systems (including plug-in electric vehicles) and

[1] Supervisory control and data acquisition.

M. Nicholson, *The Power Makers' Challenge*, Green Energy and Technology,
DOI: 10.1007/978-1-4471-2813-7_22, © Springer-Verlag London 2012

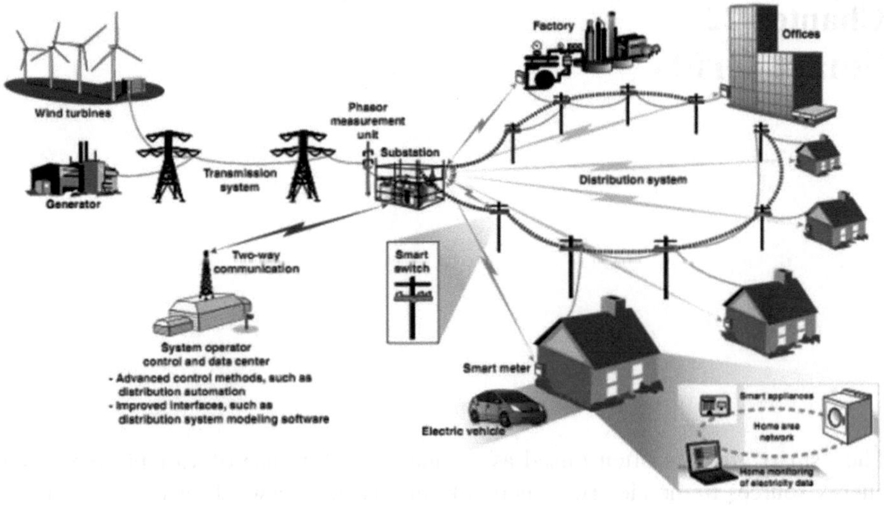

Fig. 22.1 Common smart grid components. From US Government Accountability Office [1]

automatically respond to any system disturbances or transmission congestion. It has been billed as a sort of "self-healing" network. (Fig. 22.1).

22.2 What Would be Needed for a Smart Grid?

A smart grid would need a wide-area communication network including servers and gateways as you would find in an information technology network like the Internet. It would require smarter meters that support shorter metering intervals of perhaps 5 min or less with full two-way communications, including to a home-area network to communicate with smart local thermostats and appliances with special embedded equipment.

These smarter meters would offer remote device connect/disconnect facilities to assist network reliability and advanced customer service applications such as energy management and control systems. Currently, smart meters are generally used for peak load management and are restricted to hourly intervals with limited customer services.

22.3 What are the Benefits?

We have discussed peak load management at various times throughout the book. Smart grids could optimize peak load demand through demand management and community storage as discussed in Chap. 21. This would reduce the need for additional generation and network capacity to handle the rising peak demand.

It would also reduce the average wholesale cost of electricity by shifting loads to lower cost periods (see Chap. 20).

Smart grids could make the customers more aware of their electricity usage and more able to be actively involved in what electricity they are using at any time. This could improve energy efficiency and conservation through customer energy management systems; reducing energy demand and lowering carbon emissions where the energy is coming from fossil-fuel plants. The US Department of Energy estimates smart grids could directly reduce electricity demand by 12% assuming 100% penetration of the smart grid technologies [2].

The intelligent integration of distributed storage systems will aid the assimilation into the grid of variable energy sources such as wind and solar systems. The smart grid could automatically use the storage systems to smooth out the variability. This would be particularly valuable for managing rooftop solar PV systems in local networks.

Most customer outages are caused by faults in the distribution network. With a smart grid managing the distribution network and able to reconfigure that network using automatic switching when a fault occurs, it should be possible to more quickly isolate faults and increase network reliability.

22.4 What are the Risks?

Smart technology always comes with some associated risk. The smart grid is like any other computer network relying on a secure telecommunications system. It is possible for an unauthorized person to hack into any computer network and cause problems or even serious damage. Naturally these smart grids will need comprehensive security systems to prevent any unauthorized access or cyber security threats. The consequences of such an attack could be substantial, causing widespread loss of electrical power or serious damage to equipment. Every smart meter could potentially be an access point for a cyber attack.

It will take time to retrofit existing houses and buildings with smart grid technology. It may take several decades to achieve a high penetration of smart grid technology so it will not be a quick fix for saving energy, improving reliability, or reducing energy costs.

References

1. US Government Accountability Office (2011) Electricity grid modernization. http://www.gao.gov/new.items/d11117.pdf. Accessed 13 Sept 2011
2. US DOE (2010) The smart grid: an estimation of the energy and CO_2 benefits. http://energyenvironment.pnl.gov/news/pdf/PNNL-19112_Revision_1_Final.pdf. Accessed 25 July 2011

Chapter 23
Electric Transport

As the price of oil rises, there will be more interest in using electric vehicles. Electric vehicles require battery recharging. There are several ways to recharge the batteries but most of them will require taking electricity from the grid. This will increase grid demand and load distribution throughout the day. In this chapter, we briefly discuss electric vehicles and discuss how they will be refueled. We then discuss the impact on the electricity grid which becomes yet another challenge for our Power Makers.

23.1 The Demise of Oil

Most of our transport today relies on oil. Oil will probably be the first fossil fuel in short supply. Peak Oil has been a topic of discussion for several years, and there now seems very strong evidence that we are using more oil than we are discovering which will force oil usage to decline. It will run out eventually. Burning oil-based products in vehicles produces CO_2 and other greenhouse gases which are released into the atmosphere. About 14% of world greenhouse gas emissions comes from transport. Just as we need to reduce emissions from electricity generation, we also need to do the same for transport. We need to look for cleaner fuels.

Oil will get progressively more expensive as reserves diminish. This will increase the cost of transport fuels, which will further encourage the use of replacement fuels such as NG or synthetic fuels produced using electricity. But we can also use electricity to power transport directly. Provided the electricity is "green" then the transport fuel will be green too.

23.2 The Rise of the Electric Car

Electric cars are not new. Battery electric vehicles (BEVs) have been around for over a hundred years. In the early 1900s, there were more electric cars in the US than petrol (gasoline) cars. Improvements in the internal combustion engine (ICE),

M. Nicholson, *The Power Makers' Challenge*, Green Energy and Technology,
DOI: 10.1007/978-1-4471-2813-7_23, © Springer-Verlag London 2012

and the development of widespread refueling infrastructure, favored cars using combustion engines and the BEV fell out of favor. Now it is back!

Nearly every major motor vehicle manufacturer has released, or is about to release, an electric car. Hybrid electric/petrol vehicles (HEV)-based cars (using both batteries and an ICE) have been around for several years, but the future is probably with plug-in BEVs and plug-in hybrid electric vehicles (PHEV) that use modern rechargeable battery technology to store the electricity, collectively known as plug-in electric vehicles (PEV).

BEVs produce no emissions on the road. If they are using electricity from a clean source like nuclear power or renewable energy they will produce no emissions at all.

Medium-sized trucks could also use electricity, either as BEVs or PHEVs, to reduce emissions and cut fuel costs. Delivery trucks and buses that do a lot of stopping and starting are ideal candidates for electrification because they will use significantly less energy than ICE vehicles.

23.3 How Will We Refuel Our Electric Vehicles?

PEVs are designed to be recharged in either or both of two ways. One way is to swap the battery pack when the batteries become discharged. This can be done in a few minutes at a special swap station—a bit like filling up a conventional vehicle with fuel at a fuel station. Once the batteries are swapped the depleted batteries are recharged, ready to go into another vehicle. Of course this requires an investment in suitably equipped swap stations around the country—much like the current petroleum-based fuel stations we have today.

Alternatively, the batteries can be recharged in the vehicle using a standard power outlet either at home or from a recharging point in a public car park. Recharging will take longer than battery swapping, but requires much less infrastructure. At home this may require no additional infrastructure but recharging the batteries from a standard home power outlet will take several hours.

Public car park or street recharging points (see Fig. 23.1) are likely to use high power charging outlets which could recharge batteries in minutes, depending on the battery technology.

Whether we are using battery swapping or recharging, getting the electrical energy into cars batteries will demand more grid electricity.

23.4 How Much Electricity Would Transport Need?

Electric vehicles are significantly more energy efficient than ICE vehicles. An all-electric vehicle will typically use about a third of the total energy of a similar-sized petroleum-fueled vehicle. The efficiency comes from not producing the heat generated by an ICE, not using fuel during idling and more efficient transmission and braking.

Fig. 23.1 Nissan Leaf electric vehicle and recharging station. From Wikimedia Commons Roberto and Ortiz (2010)

A typical small BEV might have a 25 kWh battery pack that will travel 150 km between recharges. So this vehicle will travel 6 km/kWh of battery energy. The energy loss during recharging will drop this to about 5 km/kWh of grid electricity used to recharge the battery. If the vehicle travels 15,000 km/year it will use around 3,000 kWh (3 MWh)/year.

An average household with two BEVs could consume an additional 6 MWh/year of electricity. This will almost double the electricity demand of some homes if all the recharging is done at home.

23.5 What Will be the Impact on the Electricity Network?

As we discussed above, if all light private vehicles switched to electricity using home charging we could see a doubling of electricity demand in residential areas. This might require a significant reinforcement of the distribution network in some places. If the charging was started after 21:00 (9 p.m.) this would keep the load out of peak time but increase the nighttime demand for electricity, traditionally supplied by baseload generators.

If there was a growing demand for fast high-powered charging the network utilities would need to provide higher power connections to many more homes than they would currently. This will almost certainly require upgrading street distribution transformers and adding to the cost of the distribution network.

Converting all transport to electricity (including trucks and buses) would very likely increase the total demand on the electricity network by at least 50%. This not only means upgrading the network but also substantially increasing the generating capacity in the network. Large long-distance haulage trucks and aircraft will probably need to use petroleum-based fuels for many decades, until on-vehicle electricity storage technology becomes practical for these types of vehicles. Alternatively, they could use synthetic fuels produced using electricity.

Electrification of transport seems inevitable. Both to replace oil consumption and to reduce transport greenhouse gas emissions. We can expect a significant increase in the demand for electricity in the future to make this transport fuel switch possible.

Chapter 24
Summing It All Up

In this final chapter we bring together the various threads involved in the Power Makers' challenge(s). We start with the carbon challenge and how we might reduce emissions. We then consider five key factors: cost, reliability, scalability, sustainability, and uncertainty, in assessing the suitability of the different technology options. We cover an all-renewable option, a clean coal and gas option, and finally a fission energy option. In practice, we are likely to use a melting pot of technologies including future technologies not yet developed.

24.1 Carbon Challenge

In Chap. 6 we discussed the carbon challenge. The bottom line was that we need to reduce our greenhouse gas emissions from electricity generation if we are to avoid unacceptable global warming. This issue will become ever more urgent as the demand for electricity grows sharply through electrification of transport and industry processes that currently rely on fossil fuels—particularly oil and gas.

We need an 80–90% reduction in emissions by 2050. Today, world electricity generation produces about 500 kg of greenhouse gas emissions for every MWh generated. This would need to be reduced to less than 100 kg, and probably to around 50 kg.

This all has to be done with the expectation that world demand for electricity will increase significantly between now and 2050—perhaps almost tripling.

24.2 Reducing Emissions

Throughout this book we have considered several ways of reducing emissions from electricity generation. They can be summarized as follows:

M. Nicholson, *The Power Makers' Challenge*, Green Energy and Technology,
DOI: 10.1007/978-1-4471-2813-7_24, © Springer-Verlag London 2012

- Reduce demand for energy by consuming less. The expected increase in world population makes this an unlikely scenario beyond a few regions. It may work for individual countries but not for the whole world. The best we can hope for is slowing down the growth in demand. Electrification of transport alone will make future electricity demand reduction practically impossible.
- Improve efficiency by using less electrical energy to achieve the same result. This can clearly help slow down the growth in electricity demand. We might reduce the energy consumed by an appliance by as much as 30% by avoiding inefficient energy use (see Chap. 5). This may not mean we will use less electricity overall because we will find more uses for electricity—such as powering our vehicles or changing industrial processes to replace fossil fuels with clean electricity.
- Reduce the emissions from electricity generation. This might be by switching to renewable energy, switching to clean coal and gas with CCS, or using more fission energy. Reducing emissions from the generating process is the path most likely to have the greatest impact on lowering total emissions.

The main thrust of this book has been to focus on the last option—switching the way we produce electricity to reduce emissions—and to think critically about the practicality of the various alternatives. Let us summarize the main options we have considered so far, paying particular attention to five key factors: cost, reliability, scalability, sustainability, and uncertainty.

24.3 An All-Renewable Future

In Part II—Renewable Energy we considered many forms of renewable energy. Some seemed more promising than others. However, all renewable resources are very "dilute" or weak sources[1] of energy and many may not be reliable, as we discussed in Chaps. 7 and 8. This means they are generally much more costly than other alternatives and can cause reliability concerns.

Renewable energy can cost 2–4 times other available alternatives (see Tables 5.1 and 20.1). These costs will be even greater if we need electricity storage, which would be essential if these were to operate on a sufficiently large scale to make a difference (Chap. 10).

Wind and solar PV can be very unreliable generators. To operate with significant quantities of energy coming from wind and solar PV, we need either

[1] We know intuitively that burning gas provides an intense source of energy, while only a modest amount of energy hits us as we sit out in the sunshine. However, there is no strictly quantitative way of making comparisons of the strengths or intensities of these different kinds of energy so instead we have to use loose subjective terms like dilute or weak. They cannot be given a precise meaning.

expensive energy storage or sufficient reliable generators to take over when the wind or sun is inadequate.

Even "reliable" renewable generators like hydro or solar thermal (Chap. 9) can be impacted by natural variability. Hydro can be affected by rainfall, and solar thermal can be impacted by extended cloud cover and substantial variations between summer and winter.

Reliable renewable generators like biomass and geothermal may not be heavily impacted by natural variability, but they may not be scalable (Chap. 9). Biomass needs plenty of arable land which may compete with food production. Conventional geothermal resources are limited around the world, and we cannot be certain about engineered geothermal (EGS) as it is not yet proven technology.

The only key factor where renewable energy scores well is sustainability—at least in fuel use. No energy conversion system can ever be fully sustainable because we still need steel and concrete to build the generators and both these building materials require non-renewable raw materials in their production—even with recycling.

A near-term, all-renewable future might seem desirable but it is not practical and it will probably be very expensive.

24.4 Clean Coal and Gas

In Part III—Clean Coal we looked at "clean" coal. Like EGS, large-scale clean coal is unproven technology. There are some serious doubts about the adequate availability and security of suitable storage sites for the CO_2.

Reliability should not be an issue for clean coal or gas. Modern coal and gas plants both can have capacity factors over 85% (see Table 5.1) which means the plants can be reliably scheduled to meet demand. If CCS can be developed commercially and scaled up economically, there is no reason to suppose that this would change.

Clean coal and gas are expected to cost 1.5–2 times other available alternatives (Table 5.1). This is still cheaper than an all-renewable option but has other disadvantages.

Sustainability is poor. The fuel resources, coal, and gas, are both limited in supply, and CCS will increase fuel use even further (see Appendix F—Carbon Capture and Storage). Proven coal reserves may be 120 years and gas is probably half of that. Any geologist will tell us that total reserves have always been much greater than proven reserves but neither coal nor gas will last forever, and electricity costs will rise as the supplies diminish. Eventually (possibly within this century), our descendants would have to find alternatives.

If suitable CO_2 storage sites prove to be a limitation then scalability and cost of coal and gas with CCS will become another issue.

There is always some uncertainty with any unproven technology. But the biggest uncertainty around clean coal and gas is the ability to store large quantities (billions of tonnes globally) of CO_2 safely without it escaping back into the atmosphere. Until this has been adequately demonstrated, a cloud will hang over clean coal.

24.5 Fission Energy

We covered fission energy in Part IV—Fission Energy. Unlike clean coal, fission energy is a proven technology that has been used for over 50 years. There are community safety concerns with nuclear power as we discussed at length in Chap. 15. It is the consensus of the scientific and engineering community that these concerns are seriously overstated. The 46 countries with reactors in operation, under construction, planned or proposed that continue to use nuclear power, clearly believe all safety issues are being well-managed.

The cost of fission energy is similar to existing technologies such as coal and gas on a levelised cost basis (i.e. over the lifetime of a plant)—see Table 5.1. Fission plants are expensive to build but in the long term, they will deliver the cheapest electricity.

Like modern coal and gas plants, fission plants have excellent reliability with high 90% capacity factors. Further, we have real-world examples of large, affluent countries, like France, which have used fission energy to effectively eliminate their reliance on fossil-fuel generation for electricity. This is a powerful vote of confidence in this technology—it has been shown to work, cost effectively, at large scales, over many decades.

Although the scalability of current generation fission plants will ultimately be limited by the availability of uranium ore, the next generation, fast reactors, could continually reuse the nuclear fuel and would be scalable long term. If fast reactors become common, the sustainability of the fuel source would no longer be an issue (Chap. 18).

Perhaps now you understand the reasons for my bold statements about fission energy in the Preface.

The uncertainty with fission energy is around the future availability of fast reactors. As stated above, we will eventually need fast reactors to make fission energy scalable and fully sustainable. Although fast reactor technology is not new, most plants of that type currently built or being built are still research reactors not full-scale commercial reactors. It may be a few decades before commercial fast reactors are commonplace. This is not necessarily a concern for fuel security as proven reserves of natural uranium are around 70 years worth and, if needed, we can use fuel recycling to extend the use of uranium. Thorium is also an alternative fuel for fission plants (Appendix G—Nuclear Power).

Of course we might not want to rely entirely on fission energy. A mix of technologies might be a more balanced solution. We still need to cover the demand throughout the day including the peak load periods. Fission plants can follow the load, but as with all generators they are most efficient at full power. It may be more cost effective to use other technologies to supply peak load and use fission plants for all the baseload.

24.6 Melting Pot

A future without fossil fuels might require a mix of fission with other efficient renewable sources. If a country has sufficient reliable hydro, geothermal, or biomass, it may make sense to continue to rely on these sources. But most countries are not in this situation.

For those countries in the sun belt,[2] solar thermal plants with adequate heat storage and backup might make sense if they ever become cost effective, although today, solar thermal is up to four times the cost of nuclear power (Table 20.1).

Eventually, wind and solar PV may be restricted to off-grid or distributed generation roles in most countries, or really only used for demand reduction. This would only change if bulk electricity storage became sufficiently inexpensive that a wind/storage combination could compete with fission, as might be the case in regions with large pumped hydro facilities.

Some see a combination of distributed generation, smart grids, and demand management (Chap. 22) as a way to move away from large centralized generators. Others suggest we need to be a less consumer-oriented society and consume substantially less energy. Neither of these solutions is realistic, and is unlikely to meet with popular appeal where the majority of developed societies seem to prefer reliable and unrestricted energy on demand. A society with a large, reliable, well-managed, centralized energy production system is much more likely to meet these demands for reliability and abundant energy.

24.7 Future Technologies

New technologies for generating electricity will emerge, and improvements will be made to existing options. Research and development continuously innovates, and we can expect all new technologies will be low-emission. Occasionally there are great breakthroughs (like fission energy) which prove to be game changers, but implementation has been historically slow [1]. Any new electricity technology will be constrained by the nature of its source of energy.

For some time, researchers have been working with algae that can convert sunlight into fuels such as hydrogen that could drive gas turbines. They may even get algae to produce electricity directly. Time will tell if algae could be such a

[2] Between 20° and 40° of latitude South and North of the equator.

game-changer, but given the physical limits on its ability to capture energy, this seems unlikely.

Nuclear fusion (as occurs in the sun) has been promising unlimited energy for decades. Big money is being spent in Europe to build a demonstration plant (see Sect. G.8), but it still seems to be decades before we will see a production facility. If it can be made to work commercially then it could definitely be a game-changer, but do not hold your breath.

Innovative storage ideas are also popping up, such as storing compressed air in bags in a lake or the ocean. Another similar idea is to use underground water storage for pumped hydro. We discussed CAES and PHS in Chap. 10. Both these ideas depend on suitable locations being available and it is not yet clear that such innovations will have a big impact on the future of electricity production.

24.8 Conclusion

We must substantially reduce emissions from electricity production if we are to meet IPCC emissions reduction targets, and those set by various individual countries. This will almost certainly demand an end to relying on fossil fuels for most of our electricity, unless CCS proves to be successful.

Replacing fossil fuels will not be easy—whatever many environmentalists might like to believe. Replacing them with a majority of renewable energy solutions will be very expensive in most countries and may even be impractical. For most countries, relying on more fission energy will not be an option—it will be an absolute necessity.

I often hear the argument that it's only the lack of political will that is stopping us from reducing emissions from electricity production. This naïve view shows a lack of understanding of the challenges facing the Power Makers. Hopefully, now you've read the book, you will have more sympathy with both the politicians and the Power Makers.

Reference

1. Smil V (2008) Moore's curse and the great energy delusion—the American. http://vaclavsmil.com/wp-content/uploads/docs/smil-article-20081119-the_American.pdf. Accessed 26 July 2011

Author Biography

Martin Nicholson studied mathematics, engineering and electrical sciences at Cambridge University in the UK and graduated with a Masters degree in 1974. He has spent most of his working life as business owner and chief executive of a number of information technology companies in Australia. His first book, *Energy in a Changing Climate*, was first published in 2009.

M. Nicholson, *The Power Makers' Challenge*, Green Energy and Technology,
DOI: 10.1007/978-1-4471-2813-7, © Springer-Verlag London 2012

Appendix A
Definition of Terms

Baseload	The minimum amount of power required to meet the expected customer demand over a standard period (such as 24 h)
Capacity credit	The 'capacity credit' or 'capacity value' of a generator is the amount of additional demand or load that can be serviced by that generator at the required reliability level
Capacity factor	The ratio of the actual energy output from a generator over a given period to the theoretical maximum output it would produce if it operated non-stop at full capacity for that period
Demand management	Controlling demand to match demand on the network to the available supply
Dispatching/Dispatchable	The planned allocation of generating plant to meet expected future loads in the network
Energy density	The energy (MJ) per kilogram (kg) or per m^3 of storage media
Electrical energy	The amount of electricity used or produced
Electrical power	The rate of consumption or production of electrical energy
Energy payback time	The time required for a system to generate the amount of energy that went into its production
Frequency control	Maintaining the network system frequency between set limits
Forced outage rate	The probability that a generating unit will not be available for service when required
Gigawatt	GW—1 billion watts or 1,000 MW
Gigawatt-hour	GWh—A GW used or produced for 1 h

M. Nicholson, *The Power Makers' Challenge*, Green Energy and Technology,
DOI: 10.1007/978-1-4471-2813-7, © Springer-Verlag London 2012

Intermediate load	Demand that is variable but predictable and influenced by time of day such as weekday mornings and evenings
Joule	International Systems unit of energy (J) defined as the amount of work done by a force of 1 N moving an object through a distance of 1 m
Kilowatt	kW—1,000 W
Kilowatt-hour	kWh—A kW used or produced for 1 h; 1 kWh = 3.6 MJ
Levelized cost	The average cost per unit of electrical energy delivered by a power station over its life-time, including capital cost, financing interest, fuel price, maintenance and operating costs
Load-following	Changing the output of a generator in response to a change in load
Loss-of-load probability	LOLP is the probability that generation will be insufficient to meet demand at some point over some specific time period
Marginal cost	The cost of producing an additional unit of electrical energy
Megajoule	MJ—1 million joules
Megawatt	MW—1 million watts or 1,000 kW
Megawatt-hour	MWh—A MW used or produced for 1 h
Merit order	Dispatching priority of generators
Negative load	Reducing the demand on the network
Negawatts	Power saved by reducing demand
Part-loaded	Generator set to provide less than full load to allow them to do load-following
Peak load	Demand that is variable but uncertain and often influenced by climatic conditions
Penetration	The total amount of energy that a generator type (e.g. wind) is supplying as a ratio of the total amount of energy generated in the network
Power density	The power produced per unit of land or water usually expressed as W/m^2
Ramp rate	The rate of change of generation usually expressed in MW/min
Reserve margin	The difference between the installed capacity, including power imported from other networks, and the maximum annual peak demand for electricity
Solar field	The area covered by solar collectors or reflectors
Solar insolation	Solar radiation per unit area usually expressed as W/m^2 or kWh/m^2 per year

Solar multiple	The ratio of the solar field to the size of the minimum solar field required to run the turbine at full capacity at noon in mid-summer
Spinning reserve	Generators that are connected to the network but are not fully loaded and can increase their power output quickly
Standing reserve	Reserve generators not connected to the network but can be connected and operating within minutes if needed
Watt	A unit of power (W) defined as a joule expended in 1 s
Watt-hour	A unit of energy (Wh) equal to 1 W of power for 1 h; 1 Wh = 3,600 J

Appendix B
Power Generation and Network Control

In this appendix we will take a more technical look at how an electricity generator works and how a utility goes about ensuring that the consumer demand for electricity is met reliably.

B.1 The Synchronous Generator

All electrical energy sources except photovoltaic (PV) use electromechanical generators to produce electricity. Synchronous generators are the most common form used for generating electricity for the grid. Wind turbines more commonly use asynchronous generators and we will discuss those in Appendix C—Wind Power.

An electrical generator converts rotating mechanical energy into electrical energy. Faraday's law of electromagnetic induction states that a conductor moving through a magnetic field will experience an induced voltage across the conductor. To generate a useful (or high) voltage it is necessary to have a strong magnetic field and a long conductor, and for the conductor to be moving as fast as possible. The best way to achieve all these is to have the mechanical-magnetic-electrical structure of the generator rotating. In practice it is preferable to have the conductor stationary and move the source of the magnetic field. It is much easier to connect a circuit or network to a stationary conductor than a moving one.

Figure B.1 shows a diagrammatic section of a simple synchronous generator with a stationary conductor around a rotating electromagnet. The conductor is known as the stator and the electromagnetic rotor has a field winding fed by a direct current (DC) source through sliding contacts (slip-rings). The field winding generates the magnetic field in the rotor. The rotor could also be a permanent magnet but is usually an electromagnet.

As the magnet rotates and the north–south pole axis reverses a variable voltage is generated in the stator conductor. With the right distribution of the stator windings the generated voltage across the stator terminals can be a sinusoidal

M. Nicholson, *The Power Makers' Challenge*, Green Energy and Technology,
DOI: 10.1007/978-1-4471-2813-7, © Springer-Verlag London 2012

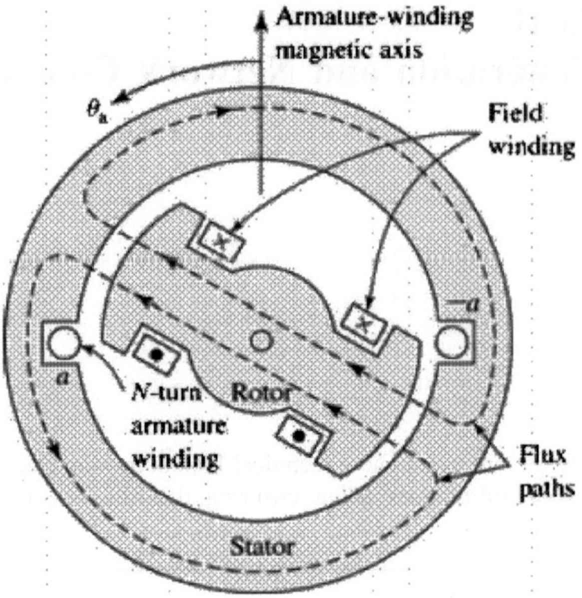

Fig. B.1 A two-pole synchronous generator. From Phuc, N [1]

Fig. B.2 Sinusoidal waveform (*AC*). From Wikimedia Commons

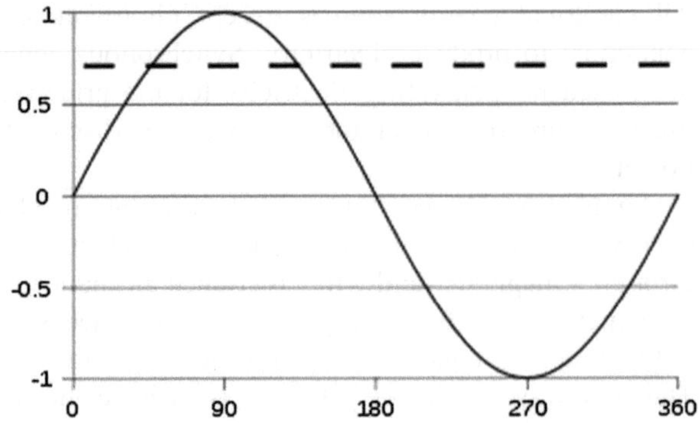

waveform as in Fig. B.2. One full rotation of the rotor will produce one complete cycle of the waveform so the frequency of the generated voltage in cycles per second is identical to the revolutions per second of the rotor. To produce a voltage frequency of 50 cycles per second with a two-pole generator it must rotate at 50 revolutions per second (3,000 rpm[1]). This is known as the synchronous speed. More poles will reduce the synchronous speed. For example a four-pole generator will have a synchronous speed of 25 revolution per second (1,500 rpm).

[1] rpm = revolutions per minute.

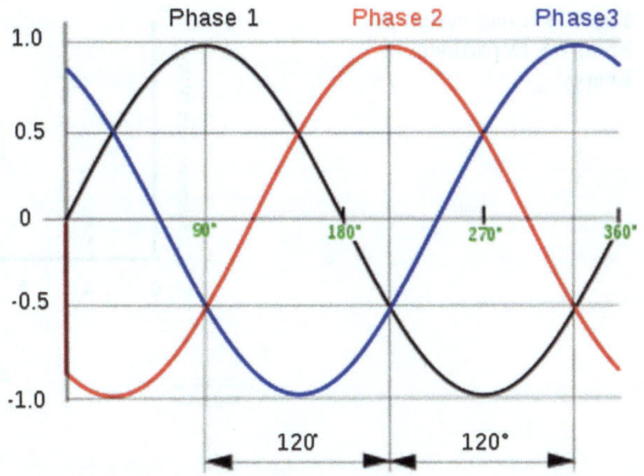

Fig. B.3 Three-phase AC. From Wikimedia Commons Messerly J (2008)

In practice, power systems use three-phase alternating current (AC) generators and transmission systems. A simple three-phase generator would use three stator windings equally positioned around the rotor axis (120° apart) so that the voltages generated across each winding are relatively shifted by 120° as shown in Fig. B.3. When the rotor is driven by an engine, for example a steam turbine, a set of balanced three-phase voltages is induced in the stator windings. When the three-phase grid is connected to the generator stator then a balanced set of currents will transfer power from the prime mover to the grid.

Before a synchronous generator can be connected to the grid it must go through a process called synchronization. The turbine first runs up the generator to its synchronous speed (see above). The DC field current in the rotor is then gradually increased until the induced voltage across the stator windings (the terminal voltage of the generator) is the same as the grid at the point of connection. Once this is done the generator can be safely connected to the grid. The generator will start to inject power into the grid once the turbine increases the torque on the generator shaft.

B.2 Power Balancing

Any increase in the load on the grid must be met by the generators or the energy stored in the system. To address this, the synchronous generator prime mover (usually a turbine) is fitted with a governor that senses any change in the speed of the generator and controls the speed by adjusting the fuel supply to the prime mover. If the load on a generator is suddenly increased it will slow down as the additional energy required is initially supplied by the rotational inertia of the generator. This slowing down will reduce the frequency. The governor will sense this and increase the fuel to the turbine to restore the speed and the frequency. As we discussed above, the generated voltage frequency is directly related to the

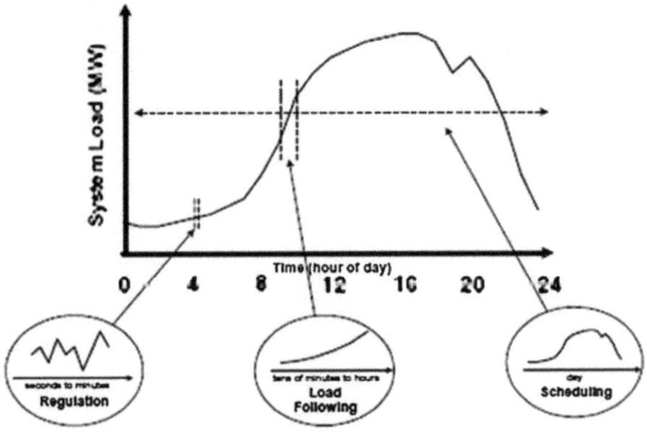

Fig. B.4 Load variations. From US Department of Energy

rotational speed of the generator, so controlling the speed will control the frequency.

The frequency in power systems needs to be kept within narrow bounds, usually within 1% either side of the norm. So for a nominal system frequency of 50 cycles per second the frequency should remain between 49.5 and 50.5. Electric motors rely on this to maintain a constant speed and transformers can also be sensitive to frequency changes. Often the actual operating band is much narrower at under 0.5% from the norm. Significant drift in the system frequency of more than say 6% could have dramatic effects on network stability and damage connected equipment.

The governor on a specific generator is set at a particular frequency. This 'set point' is the system frequency at which the generator will supply no power. As the system frequency starts to fall (due to increased demand on the grid slowing down the generators), then the governor will increase the fuel supply and the generator will start to produce more power to meet the demand. Where the full rated output of the generator is to be used then the set point would be adjusted so that full power is reached when the system frequency reaches the nominal frequency. This might be the case for the least cost generators (in dollars per additional MWh) to ensure that all their power is normally used. If the system frequency continued to fall below the nominal because of further demand on the grid then these generators would be unable to contribute further power to meet this increased demand and the power system would need to rely on other generators that are not fully loaded (i.e. part-loaded).

The system frequency and the set point of the governor are used to control how much demand is placed on each generator in the system. In practice, the closer the set point is to the nominal frequency the less power it will supply once the nominal frequency is reached but there will still be spare capacity to supply further demand if needed. This is called part-loaded. Typically, the higher running cost plants will usually be run part-loaded (for example gas OCGT[2]) and the lowest cost plants will be run fully loaded (e.g. nuclear, coal, or gas CCGT[3]).

---

[2] Open Cycle Gas Turbine.

[3] Combined Cycle Gas Turbine.

B.3 Demand Matching

Figure 4.1 shows the daily variation of a typical load on a power system. The load is lower at night and starts to peak in the morning and afternoon. The shape of the load curve often varies between summer and winter and will change depending on the climate that the power system operates in. Not shown on that curve are the continuous smaller and faster random variations in load as consumers turn on and off equipment.

These short duration variations in load of seconds and minutes (see Fig. B.4) are addressed with part-loaded plant under governor control as described above. These plants are generally large generators and are said to be operating in "frequency response" mode—referred to as "Regulation" in Fig. B.4.

For longer duration variations of several minutes to several hours, plants are connected or disconnected to balance anticipated load increases or decreases. These plants are said to be "load-following" and typically supply the "intermediate load" shown in Fig. 4.1. An example of such relatively slow and predictable changes in demand would be the morning increase around 06:00 (6 a.m.) also shown in Fig. B.4. This relatively slow change can be impacted by significant quantities of wind power operating as negative load (see Chap. 8).

Fission plants are generally less flexible and best operated at a steady load although this is changing with newer generation fission plants. This makes them best suited to the valuable role of providing the baseload power shown in Fig. 4.1 . Coal, CCGT, and hydro with reservoir are more flexible and can be used for both baseload and load-following as can some newer fission plants. Gas-heated steam turbines are also used for load-following. OCGT and pumped storage hydro are usually used at peak times and hydro plants can also be used to handle peak load.

B.4 Demand Forecasting

Scheduling plant to meet the anticipated load requires an accurate prediction of future demand through the day. Scheduling is the responsibility of the system operator who coordinates the dispatch of generating units. Steam turbines can take several hours to prepare for connection to the grid so they need to be scheduled well ahead of time. Sometimes they are left running part-loaded and ramped up as needed. Gas and hydro turbines can be connected and disconnected from the grid much more quickly than steam turbines.

Because demand tends to have regular patterns this forecasting can be reasonably accurate up to 24 h ahead—often within 3%. Occasionally the load may exceed the forecast so it requires reserve capacity to maintain system reliability as discussed in Chap. 4.

B.5 Frequency Control

In Sect. B.2 we discussed the need for the frequency to be kept within narrow bounds. This requires a response from generators over different time periods. Fluctuations in frequency caused by modest changes in demand can be controlled by the frequency response generators discussed above.

An abnormal drop in frequency caused by the sudden loss of a generator or a significant increase in demand will initially be controlled by the governor action of large part-loaded generators that can start to redress the lost balance within seconds. This is the spinning reserve we discussed in Chap. 4. If this reserve is unable to replace all the lost power within about 30 s then further spinning reserves will be called on which may be slower to respond but can replace the power long enough to bring additional standing reserves online if needed. Standing reserves can be OCGT or hydro plants that need to be ramped up to synchronous speed and connected to the grid within 10–20 min.

Some networks have specific customer loads (often industrial) that operate under frequency control. The loads are often called "interruptible" loads that can be instantly disconnected to reduce the demand when needed. These dropped loads act like spinning reserve to restore the frequency within the required limits. This is often referred to as "demand management" or "demand response."

In extreme situations, loss of generation can be controlled by load shedding. Networks have groups of non-critical loads that are automatically switched off the network at different system frequencies. The least critical are switched off first but if the system frequency continues to fall then more loads are shed until the frequency stabilizes.

A sudden loss of demand—often caused by a transmission line failure perhaps from a lightning strike or tree fall—may cause an abnormal increase in system frequency. Again this will initially be controlled by governor action which will reduce the power output from the generators. If the loss is so great that the frequency rises significantly then generators will be automatically tripped off the system until the power balance is reinstated.

B.6 Dynamic Demand Control

So far we have discussed how system frequency responds to the load or demand. This relationship between the power balance and the system frequency can be used to automatically control demand. We briefly discussed the process of load shedding above whereby a drop in system frequency can switch off sections of the network to restore the balance between demand and supply.

With dynamic demand control (DDC), individual loads (such as water heaters, refrigerators, or air conditioners) can be fitted with a small electronic controller that responds to the system frequency and switch the appliance on or off automatically for short periods of time without significant impact on the appliance user.

DDC could also be used on domestic deep-freezers, air-conditioners, pumps, and water heaters as well as industrial and commercial units. There are literally millions of such appliances in most electricity networks. A significant number of DDC loads can smooth out demand in the network and improve the frequency stability. This will reduce governor activity on generators and possibly the need for some reserve capacity. It will also help with managing variable RE generation.

B.7 RE Impact on Network Control

Variable RE generators have both a balancing impact and a reliability impact on a network. With greater use of variable RE generators in the network, power balancing and frequency control will become more challenging. At modest levels of RE penetration in the network these challenges are manageable but might require additional reserves, these need to be considered when valuing electricity from RE resources.

An indication of the impact of variable RE resources can be found in Freris and Infield's book *Renewable Energy in Power Systems* [2] where the authors looked at the impact of wind power penetration in Europe. They suggest that, for up to 5% of the electrical energy coming from wind power, no special measures are required. Between 5 and 10% occasionally some wind energy would need to be discarded and more spinning reserve would be needed. Above 10%, more storage will be required or more conventional generation will need to be retained. Also the cost penalty from the variability rises with increased penetration. We will discuss this in more detail in Appendix C—Wind Power.

The extent of the impact will increase with penetration and vary from network to network. It can depend on the quality of the RE resources, the accuracy of forecasting these resources, the amount of storage in the network, the strength of the grid, interconnection to other networks, and the mix of generators used.

Some RE generation can contribute to frequency response services but, as with traditional fossil fuel generators in the role, they would need to be part-loaded. Biomass and hydro can perform this role, but so can modern large-scale solar PV using power electronic control systems and variable speed wind turbines with blade pitch control (see Appendix C—Wind Power).

Appendix C
Wind Power

This appendix takes an inside look at wind power, the type of generators used, and how the power output relates to wind variability. It also looks at various schemes aimed at coping with variability.

C.1 Wind Variability

The wind speed varies continuously at a given location. It can vary from second to second (called "turbulence"), from day to day (called "diurnal"), with passing weather patterns (called "synoptic") and with changes of seasons and changes from year to year.

The overall energy captured from the wind over the year is controlled by the annual and seasonal changes. The impact on electricity networks from wind power is more from the synoptic and diurnal changes than from turbulence, as we will discuss later.

The longer term variations of wind speed can be shown using a frequency or probability distribution. Figure C.2 shows a typical wind speed frequency distribution graph showing the speed in meters per second (m/s) and the probability or frequency of getting that wind speed at a particular site. The mean wind speed in this example is just under 7 m/s (25 km/h) and occurs 10% of the time. The long tail off to the right reflects that very high winds do occur occasionally but infrequently. Not all sites would have such a high mean wind speed but 7 m/s is thought to be the required profile for a "good" site. Some would argue this is the minimum for commercially viable wind power.

C.2 Power Variability with Wind Speed

Wind variability impacts the power output of wind generators. The energy that can be extracted from the wind varies as the cube of the wind speed so if the wind speed doubles then the kinetic energy of the moving air increases eightfold [2^3 (cubed) $= 8$].

M. Nicholson, *The Power Makers' Challenge*, Green Energy and Technology,
DOI: 10.1007/978-1-4471-2813-7, © Springer-Verlag London 2012

Fig. C.1 Windmills and sheep in the Netherlands. From van Urk T (2009)

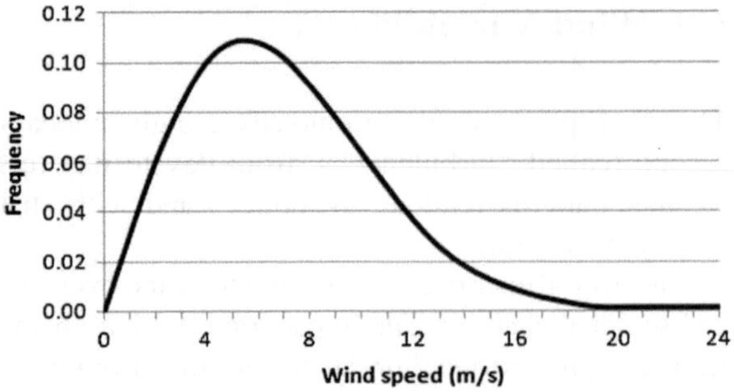

Fig. C.2 Wind speed variation

The height above the ground impacts the wind speed and hence the power that can be extracted from the wind. Wind speed increases with height in a complicated way and depends on the roughness of the surrounding terrain. Roughly speaking, doubling the height increases the wind speed by 10% and therefore the power of the wind by 30% [3]. So the taller the wind turbine, the more power can be extracted.

At low to medium wind speeds (below 13 m/s) the electrical output from a single turbine can vary substantially as can be seen from the power curve in Fig. C.3. If the wind increase from 11 m/s (40 km/h) to 14 m/s (50 km/h) the power output doubles.

During a wind lull below 5 m/s (18 km/h) there will be very little, if any, power generated by the turbine.

Above 13 m/s it is possible to maintain a constant output from a wind generator, as we will discuss later in Sect. C.5. This certainly helps dealing with wind variability in the network, but as we saw from Fig. C.2, most of the time the wind speed is below 13 m/s.

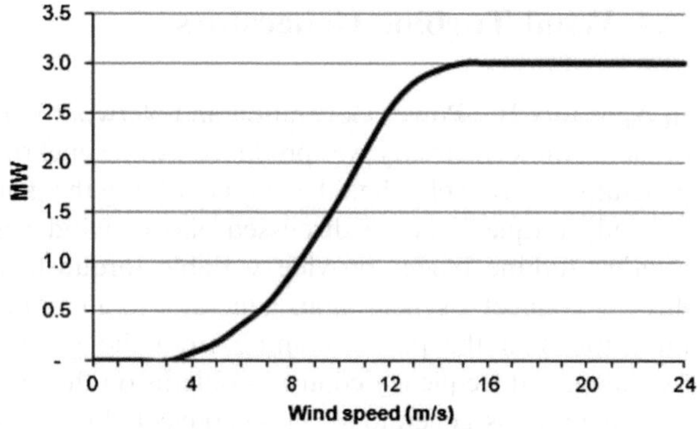

Fig. C.3 Wind turbine power curve

Above 25 m/s (90 km/h)—known as the "cut-out" speed—wind turbines shut down to avoid possible mechanical damage to the turbine and generator.

C.3 Wind Farms

Wind turbines are commonly clustered in wind farms and sufficiently spaced such that the turbulence at each turbine is different. Typically, turbines are spaced at least five times the diameter of the blades apart, to minimize the impact on each other of wind-shadows[4]—which means that modern turbines need to be about 500 m apart. As the output from each turbine is added together the overall impact of turbulence on the wind farm is reduced and the variability of electrical output due to turbulence is less. Also, spinning turbines have inertia, so second-by-second variations in wind speed do not necessarily change the turbine shaft speed greatly. The output variations from minute to minute for a wind farm will not be excessive. Of course a lull that affects the whole wind farm will mean little or no power from that farm.

Wind farms vary in size, with an average power rating of around 120 MW. The largest wind farm so far has over 400 turbines totalling 735 MW in Texas. Wind farms are typically left unattended by operators but are monitored and controlled by a computerized supervisory control system.

A region or state will often have several wind farms geographically separated. The variation in output from each farm may vary widely from hour to hour, but an aggregation of the electrical output from all these farms will show less variation. However, wide area synoptic wind patterns do happen across a whole region impacting the total amount of wind power available (see Fig. 8.1).

[4] A wind-shadow is where a turbine reduces the wind speed for a downwind turbine.

C.4 Wind Turbine Generators

In Appendix B—Power Generation and Network Control, we discussed how most of the world's electricity was produced using synchronous generators. Synchronous generators work well when the engine driving the generator provides what is called a steady torque.[5] As we discussed above, wind varies continuously and so the rotating turbine blades provide variable torque to the generator. If we were to directly connect a synchronous generator to a wind turbine it would produce large fluctuations in the power coming from the generator which can cause power balancing and frequency control problems on the grid.

Synchronous generators can be connected to a wind turbine through a gearbox to partially compensate for the variable turbine speed. Today these only tend to be used in stand-alone networks that are not directly connected to the grid, often in remote communities.

Synchronous generators can be used with grid connected wind turbines, with or without a gearbox, by using power electronics to synchronize the output to the grid frequency. However, most modern wind turbines use asynchronous generators.

Asynchronous or induction machines (which are largely motors) are the most widely used electro-mechanical devices. They are used to power industrial machines, pump water, and are even found in household appliances like fans and electric drills. Asynchronous machines can be both motors and generators.

Asynchronous machines work in much the same way as synchronous generators, except that they do not need an external current fed to the rotor coil. An external AC is fed to the stator which generates a rotating magnetic field. This rotating magnetic field from the stator induces a current in the rotor, creating its own magnetic field, and causing the rotor to rotate. This is how an induction motor works. If the rotor is then turned by an engine (e.g. a wind turbine) faster than the rotating field an AC is generated in the stator and the machine acts like a generator.

The big advantage of asynchronous generators for wind power is that they are less expensive and more robust than synchronous generators. They can produce useful power for the grid at variable rotor speeds using power electronic converters.

Wind turbines will often use four pole asynchronous generators with a synchronous speed of 1,500 rpm, rather than the 3,000 rpm normally used in synchronous generators in large power stations (50 cycles per second). The wind turbine needs to turn the generator rotor slightly faster than 1,500 rpm to generate electricity. Given the size of wind turbine blades (50 m) the tip speed of the blade would need to be travelling very fast to rotate the turbine shaft at 1,500 rpm. Modern turbine blades rotate at about 10–20 rpm with a gearbox between the turbine and the generator rotor shaft.

Various mechanisms are used to reduce the impact of variations in wind speed with variation in power production from the generator. The turbine blade pitch can be adjusted to slow down or speed up the turbine for a given wind speed, which takes

[5] Torque is the tendency of a force to cause or change rotational motion.

care of longer variations in wind speed. Faster variations in wind speed caused by wind gusts can be handled using integrated electronics on the rotor shaft to control the speed of the generator. This is the process used to generate the power curve shown in Fig. C.3.

C.5 Wind Network Control

Modern wind turbines can also make a contribution to network system control, including grid voltage control, power balancing, and frequency response discussed in Appendix B—Power Generation and Network Control. These are achieved by electronically controlling the electrical output from the wind turbine generator [4].

C.6 Coping with Wind Variability

As we have discussed already, the wind speed varies continuously. Some of this variation, from seconds to minutes, can be handled using modern turbine technology, but long-lasting periods with little or no wind called "lulls", and short-term changes in wind farm output, called "slews" or "ramps", can still cause management issues on the network.

Ramps in a network can be on both the supply and the demand side of the power balancing equation. Network operators today already cope with ramps in demand by changing the supply from conventional generators (see Sect. 4.1). In theory, wind slews could also be managed with sufficient conventional generators able to ramp up or down to maintain the power balance.

Total wind power variability across a given region or state can be large over several hours and days, even with geographically distributed wind power as can be seen in Fig. 8.1. The figure shows the total network load over 21 days in Western Denmark and the corresponding total wind production over the same period from all the wind farms in the region. Sometimes the wind power fell sharply to below a quarter of the peak for several lengthy periods. On one occasion, 5 days into the period, a storm caused the wind power to fall from a peak of 2,100 MW to less than 10% (around 150 MW) in around 6 h. If Denmark was to rely only on its own conventional generators to provide the lost power over that 6-h period it would need to significantly increase its reserve capacity to cope with this kind of ramp. Fortunately, Denmark can buy and sell power from and to its neighbors when needed so it is better able to cope with such ramps.

Many countries, such as Australia, do not have access to their neighbor's power when needed. This makes it very difficult for these countries to use significant amounts of wind energy and maintain reliable networks. The generally accepted upper limit is around 25% of total energy demand [4].

Even with moderate amounts of wind power, we need to be able to forecast these lulls and slews several hours to a day ahead, to give the network operators time to schedule conventional power plants to manage the anticipated changes in wind power.

C.7 Wind Forecasting

In large-scale distributed wind power systems variability increases as the timescale increases. The second and minute variability across distributed wind farms is generally small, and can be handled by modern wind power technology but the variability over several hours can be large, as we have seen, and needs to be forecast where possible.

The objective of a wind forecasting system is to provide predictions of wind power output for certain time horizons (the warning time) with a suitable frequency (for example hourly predictions), taking into account the wind speed and direction and the local terrain in the area of the wind farm. These wind forecasting tools use the wind predictions to enable the system operator to anticipate the power output from the installed wind capacity.

For short-term forecasts of under 6 h, wind predictions are performed using statistical methods based on observations and available meteorological data. These meteorological data are typically provided by meteorological bureaus.

For a time horizon of over 6 h to around 4 days, wind forecasts rely on numerical weather predictions (NWP) systems (complex computer simulation models) that provide wind speed and direction forecasts that can take several hours to prepare.

As we discussed in Chap. 4, short-term load forecasting can be done with some accuracy 24 h ahead. Forecasting wind power over the same timescale is more difficult. Large deviations between the forecast and what actually happens can occur in both the strength and timing of winds and therefore in wind power generation. Denmark prepares wind power forecasts 13–47 h ahead, with an average error of 30–35% [5].

This uncertainty can be an issue for network system operators tasked with scheduling generating capacity. Combining forecasts over a larger area can improve the level of accuracy, as can forecasting over a shorter time period.

C.8 Wind Backup Needed for Reliability

If wind power is to be a significant proportion of a network's generating capacity then the network system operator has to be able to cope with the loss of some, or most, of that wind power for a period of time which could be hours to days.

In the main body of the book we discussed reserve margin, capacity factor, and capacity credit. We will now bring these three things together to calculate how much conventional capacity can be replaced by wind power while maintaining network reliability.

The capacity credit of an RE source was defined as the amount of conventional fossil-fuel plant that can be replaced and maintain the same system reliability (see Sect. 11.3). System reliability is often expressed in terms of the loss-of-load probability (LOLP). LOLP is the probability that generation will be insufficient to meet demand at some point over some specific time period. Or to put it another way, LOLP is the probability that a loss-of-load event (LOLE) occurs whereby the demand on the network is not met. A system operator might aim for only one such major event to happen every 10 years.

To maintain the required level of reliability, the network operator requires a reserve margin to be maintained, as discussed in Sect. 4.3. For example, let us take a network with a peak demand of say 50 GW using conventional generators. This might require a reserve margin of 10 GW (20% of the peak), so the total installed capacity would actually be 60 GW.

Let us now consider that 10% of the system energy in this example was to be generated by wind power and let us assume that this system has a yearly energy demand of 250 TWh. Ten percent of this demand would be 25 TWh. At a capacity factor of 30% we would need 9.5 GW[6] of wind capacity to generate 25 TWh. Using a similar calculation, if we used conventional generators for the 25 TWh at a capacity factor of 75%, we would need 3.8 GW of conventional generators. Ideally, the 9.5 GW of wind capacity would replace all the 3.8 GW of conventional capacity, but because of the variability of the wind we cannot replace all the conventional capacity and maintain the same LOLP.

The network operator might statistically calculate that if 9.5 GW of wind capacity was used in the network then the installed conventional capacity would need to be 58 GW rather than the 60 GW previously required. So the 9.5 GW of wind capacity has only removed 2 GW of conventional generation from the original network. This is the capacity credit of the wind capacity or 21% of the installed wind capacity.

1.8 GW (3.8–2 GW) of the conventional capacity needs to be retained to cope with the variability of the wind power. This can be considered to be the backup generation needed for the wind power so in this example the backup needed would be 19% of the installed wind capacity.

C.9 Cost of Wind Integration into the Network

The need for backup generation comes at a cost, and forms part of the cost of integrating wind into existing grids. But there are additional costs such as the cost of managing the wind variability using wind forecasting and the need for additional balancing capacity. Wind farms are often located away from load centres and this requires additional transmission costs. These integration costs per MWh increase with wind penetration.

To the system operator, network security and reliability is paramount and wind variability creates additional costs to maintain that reliability. What are referred to as the 'balancing and capacity costs' (to maintain this reliability) make up the largest component of wind integration costs (yet are rarely included in typical LCOE estimates). Capacity cost is the cost of the additional backup capacity

---

[6] 9.5 GW = 25,000 GWh/(365 days × 24 h × 0.3 capacity factor).

(which could be additional OCGT standing reserve capacity) needed for reliability as discussed above. Balancing costs include the additional synchronized reserve needed for power balancing (see Sect. B.2) plus the wind energy curtailment costs during periods of low demand (see Chap. 8).

Smaller wind farms are often connected to the distribution network, whereas larger wind farms may be connected to the transmission network (see Chap. 3). In both cases the wind farms are often away from the existing network and require existing network reinforcement to handle additional capacity and extension to reach new locations with the best wind resources. Additional transmission costs include the costs of reinforcement, line extension and power losses in transmission.

Modeling estimates done by ILEX Energy Consulting [6] showed that the total integration cost for 20% penetration of wind in the UK was around US $18/MWh and comprised 48% capacity costs, 23% balancing costs, and 29% transmission and distribution costs. Based on costs in Table 5.1 this adds about 10% to the LCOE cost for wind.

Appendix D
Solar Power

Solar electricity generators comes in two flavors—PV and solar thermal, sometimes referred to as concentrated solar power (CSP) or solar thermal electric (STE). In this appendix we review the technology used in solar systems as well as important attributes of the various technologies when producing electricity.

D.1 Solar PV Systems Technology

There are two different technologies used to make solar PV systems: crystalline silicon (Si) and thin-film.

Today, 85% of PV panels are made using crystalline Si. These panels can be single-crystal or multi-crystalline. Single-crystal panels are generally more expensive to manufacture than multi-crystalline but are more efficient. A crystalline Si panel typically consists of a number of Si cells connected together using a metal grid. A single cell only produces 1 or 2 W so a number of cells are connected so that a single rigid panel or module can produce several to hundreds of watts.

Thin-film is newer technology and reduces the amount of light absorbing material required to make the cell. As a result it is less expensive to make but also less efficient. Thin-film uses different semiconductor materials such as copper indium diselenide (CuInSe2) or cadmium telluride (CdTe) and the cells are flexible and can be used in various building applications such as roof tiles. These materials are rarer and more expensive than Si and more difficult to dispose of safely.

Both crystalline Si and thin-film cells work best at lower temperatures. As the operating temperature rises the efficiency falls.

The output from PV cells can be increased by using concentrators to capture more solar energy and focus it on to the cell. The concentrators could be lenses or reflectors and the systems can also be designed to track the sun. The advantage of concentrators is more power output from each PV cell. Multi-junction Gallium

M. Nicholson, *The Power Makers' Challenge*, Green Energy and Technology,
DOI: 10.1007/978-1-4471-2813-7, © Springer-Verlag London 2012

Table D.1 Efficiency and energy payback time for PV roof panels

Technology	Typical conversion efficiency (%)	EPBT years[a]
Single-crystal Si	14	2.7
Multi-crystalline Si	12	2.2
Thin-film CdTe	9	1.0

[a] From Alsema et al. [9]

arsenide (GaAs) concentrator cells can achieve concentration ratios exceeding 400 times and efficiencies exceeding 40% [7]. The disadvantage is the additional cost of the lenses or mirrors and tracking systems which can exceed the cost savings from concentration. The cells also need to be able to function at much high temperatures from the concentrated sunlight.

In general, concentrated PV is more expensive per kWh than standard PV panels and uses special high-efficiency multi-junction cells to avoid the temperature degradation seen with crystalline Si.

D.2 Solar PV Installations

Small single PV panels are ideal for small-scale stand-alone applications such as powering roadside equipment like phones or parking meters. At the next level, PV panels are normally installed in clusters with, for example, several panels on a domestic roof. In a solar PV farm there may be many hundreds of panels.

All PV panels generate DC. For stand-alone applications this is generally adequate but to connect PV panels to the electricity network it is necessary to convert the DC to AC using an inverter. Several panels can be connected to a single inverter but some panels use an integrated mini-inverter to produce an AC output.

Panels are typically installed at a fixed tilt and direction to absorb the maximum amount of sunlight at that location. PV panels can also be mounted on tracking mechanisms to allow the panel to face the sun as closely as possible throughout the day, which increases the radiation falling on the panels.

D.3 Solar PV Conversion Efficiency

The conversion efficiency of a PV cell is the proportion of sunlight energy that the cell converts to electrical energy. As we saw above, the conversion efficiency depends on the technology used to make the cell. Table D.1 shows typical conversion efficiencies for PV roof panels.

D.4 Solar PV Power Density

The power output from a PV system depends on the actual amount of sunlight falling on the panels often referred to as "solar insolation" and the conversion efficiency of the system.

The radiation that reaches sea level at high noon in a clear sky has an insolation of about 1,000 W/m^2. A solar panel with a 10% conversion efficiency at that location would deliver 10% of 1,000 W/m^2—a power density of 100 W/m^2. This is the peak density, so the average over the day would be much less depending on geographic location and season—perhaps 5–20 W/m^2 [8]. As a result, this panel might generate 0.5 kWh/m^2 on a bright sunny day in the right geographical location.

D.5 Solar PV Energy Payback Time

It takes energy to make solar PV systems. So how many years does it take before the PV system generates more energy than was used to make it? The energy payback time (EPBT) is the length of time required for a PV system to generate the amount of energy equal to the total energy that went into its production. Table D.1 shows the EPBT for PV panels.

D.6 Solar PV Power Variability

Solar PV power generation is directly related to the solar insolation on the solar cells. Like wind, solar radiation, and hence insolation, varies. Unlike wind, solar radiation is never available at night so the variability is limited to between sunrise and sunset. Changes in insolation are caused by the elevation of the sun, cloud cover, moisture content in the air and other impurities such as dust particles.

Figure D.1 shows an example of minute-by-minute variation in solar insolation over a 12-h period. Clearly, the actual shape of the graph will vary between locations and depends on the cloud cover. The sudden dips in insolation are caused by clouds blocking the sun's radiation. The degree of variation depends on the density of the clouds and the cloud passage time.

These sudden dips in insolation can cause significant changes or ramps in the power output from a solar PV system over a few seconds. These short-term variations are much more prevalent with solar PV than with wind power or solar thermal systems because there is no inertia in a solar PV panel as there is in a wind turbine (see Appendix C—Wind Power). As a result, solar PV can have a potentially large impact on load fluctuation in the network, with a corresponding increase in demand for frequency response services (see Appendix B—Power Generation and Network Control).

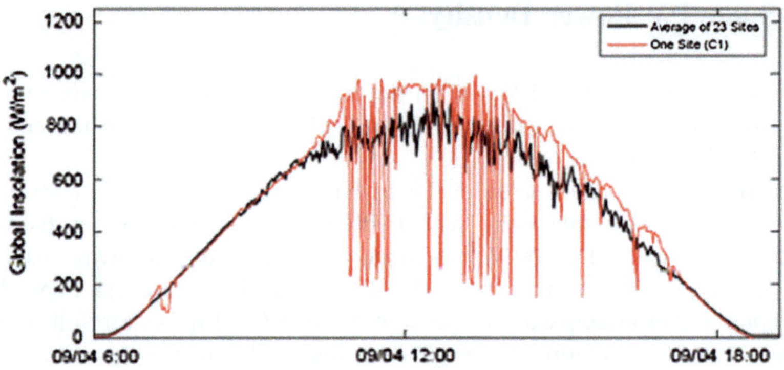

Fig. D.1 Solar Insolation. From Mills A (2009) Lawrence Berkeley National Laboratory [10]

Ramps in solar PV output are particularly prevalent in the morning and afternoon, as can be seen from Fig. D.1 and can have some impact on load-following services in the network.

Figure D.1 shows the solar insolation at a single site (the thinner line) and the aggregated insolation from 23 sites (the thicker line) spread over 400 kms. As with wind, geographic separation of a number of solar PV farms should reduce the variability of power output seen from just one site.

D.7 Solar Forecasting

Insolation forecasting is relatively new and building on the experience already gained in wind forecasting (see Appendix C—Wind Power). Short-term forecasting over the next few minutes is performed using sky cameras. The visual images from these cameras can be used to predict cloud movements over a particular location and calculate the impact on solar PV output.

Similar medium-term forecasts over several hours are developed from satellite imagery.

D.8 Solar Thermal System Technology

Solar thermal systems used for generating electricity first convert sunlight to thermal energy (heat) and then use the thermal energy to drive a turbine/generator. All CSP systems use mirrors to concentrate the sunlight onto a receiver where it is converted into heat. Mirrors can concentrate the sunlight up to 1,000 times. The best locations for CSP systems is between the 20th and 40th degree of latitude south and north of the equator—often referred to as the "sun belt".

There are three technologies in use to collect and concentrate the sunlight and convert it into usable thermal energy to generate electricity: linear concentrator systems, power tower systems, and dish/engine systems.

Fig. D.2 Parabolic trough collectors in the Mojave Desert USA. From Orlemann A (2009)

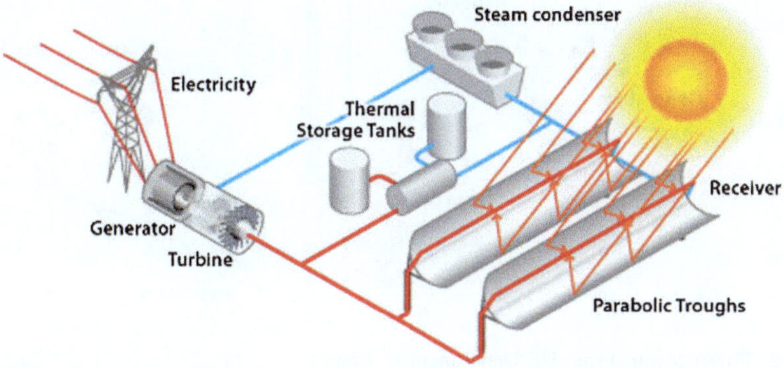

Fig. D.3 Parabolic trough collectors. From the US Department of Energy

Linear concentrator systems use large, long mirrors to focus the sunlight onto a linear receiver tube containing a heat-transfer fluid (see Fig. D.2). The fluid can be oil, molten-salt, or even water/steam. The heated fluid is transferred to a heat exchanger where it is converted into steam to drive a steam turbine. If the fluid is water the steam generated can be used directly in the turbine. The collector mirrors can be in the shape of a parabolic trough (Fig. D.1) or flat or slightly curved known as Fresnel reflectors. Fresnel reflectors are cheaper, but less efficient, than parabolic troughs.

In parabolic trough systems the receivers are at the focal line of the parabola-shaped trough as shown in Fig. D.3. Fresnel reflectors are mounted on trackers and

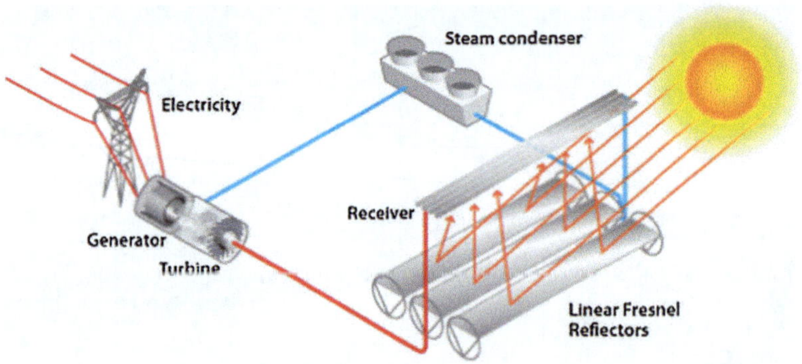

Fig. D.4 Fresnel reflectors. From US Department of Energy

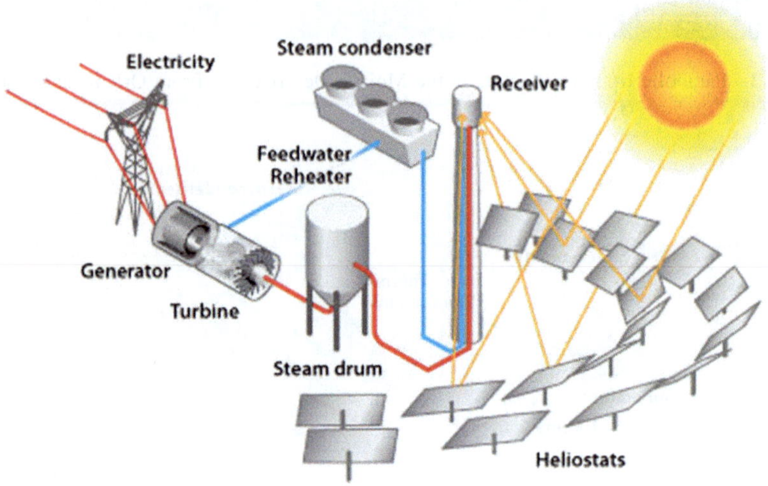

Fig. D.5 Power tower. From US Department of Energy

reflect the sunlight onto a receiver fixed in space above the mirrors as shown in Fig. D.4.

With power towers the receiver is at the top of a tall tower. A large number of flat mirrors called heliostats track the sun and focus the sunlight onto the central tower receiver, as shown in Fig. D.5. As with linear concentrators, the receiver can contain a heat-transfer fluid that is converted to steam in a steam generator to drive a steam turbine. Tower receivers can generate high temperatures which makes it possible to also generate hot air for a gas turbine which can be used in an integrated solar combined-cycle turbine (ISCC).

Linear concentrators and power towers have an advantage over other solar systems in that the heat collected can be stored for use when the sun is not shining. We will discuss solar thermal storage shortly.

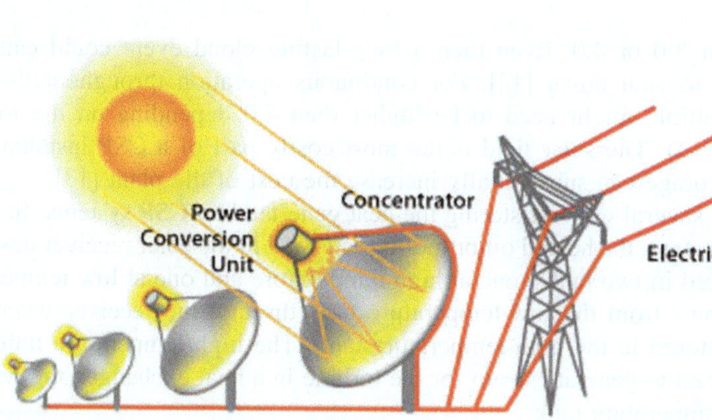

Fig. D.6 Dish/engine. From US Department of Energy

Dish/engine systems use a large solar concentrating dish to collect the sunlight and focus it onto a power conversion unit, as shown in Fig. D.6. The power conversion unit consists of a thermal receiver that collects the solar thermal energy and heats a fluid that drives a heat engine and generator. The most common heat engine used in dish/engine systems is the Stirling engine. A Stirling engine uses the heated fluid from the receiver to move pistons and create rotating kinetic energy to turn a generator. Individual dish/engine systems are generally small generators under 25 kW [11].

D.9 Solar Field and Solar Multiple

The solar field is the area covered by solar collectors or reflectors. The minimum solar field required to run the turbine at full capacity at noon in mid-summer is defined as a solar multiple of 1.0. This is often referred to as the "design point".

To extend the use of the turbine at full capacity into hours of reduced solar insolation, it is necessary to increase the size of solar field by a factor which is known as the solar multiple. A solar field with a multiple of 2.0 is twice the size of the design point field. The field in a production plant might have a solar multiple of 1.5–4.0.

D.10 Solar Thermal Storage

Storing the heat from CSP systems can significantly increase the capacity factor of the turbine/generator (often referred to as the power block) by extending the period of operation. The stored heat can be used after the sun has set or during extended cloudy periods.

Storage requires increasing the solar field to provide the extra energy to be stored. To use a CSP plant continuously in summer might require a solar field with a multiple of 3.0 or 4.0. Even then a long-lasting cloud event could cause the power plant to shut down [12]. For continuous operation throughout the year, the solar multiple might need to be higher than 4.0 depending on the location (see Sect. D.13). The solar field is the most costly part of a CSP installation so significant storage can substantially increase the cost of the plant [13].

There are several ways of storing the heat generated by CSP systems. In a two-tank direct system, the heated oil or molten salt used in the solar receiver described above is stored in two tanks, one at high temperature and one at low temperature. The fluid flows from the low-temperature tank through the receiver where it is heated and stored in the high-temperature tank. The high-temperature fluid from the tank is used to generate steam for the turbine in a heat exchanger and returned to the low-temperature tank.

In a two-tank indirect system, different fluids are used in the solar receiver and the storage tanks. This is done by heating the fluid from the low-temperature tank in a heat exchanger using the hot heat-transfer fluid coming from the solar collectors. Once heated it flows back to the high-temperature tank. Typically, in an indirect system, the solar receiver fluid would be oil and the storage fluid would be molten-salt.

A single-tank systems can also be used where the heat is stored in a solid medium such as silica sand where the top of the tank is at high temperature and the bottom of the tank at low temperature. The hot heat-transfer fluid from the solar collectors flows in at the top and out at the bottom at low temperature, after having heated the solid medium. To extract the heat from the storage the process is reversed and the cool heat-transfer fluid enters at the bottom and exits at the top after being heated.

Extensive research is being performed to improve heat-transfer fluids and storage medium with a view to increasing efficiency and reducing storage costs [13].

D.11 Solar Thermal Co-Firing

As well as storing the heat, it is also possible to use chemical fuels such as oil, gas, coal, and biomass to provide additional heat to extend the use of the solar thermal power block when the solar-generated heat and the stored heat are exhausted. Solar thermal plants that use additional sources of heating are often referred to as "hybrid" plants.

Solar collectors can also be used to "co-fire" a conventional thermal power station by providing additional heat to the steam generator. This can reduce the use of coal or gas fuels.

Table D.2 Solar thermal conversion efficiencies

Technology	Solar field efficiency (%)	Power block efficiency (%)	Solar to electricity efficiency (%)
Parabolic trough	43–56	34	15–19
Fresnel reflector	34	35	12
Power tower	46–53	34	16–18

From NEEDS [13]

D.12 Solar Thermal Conversion Efficiency

With CSP systems we need to consider both the solar-to-heat efficiency within the collector and the heat-to-electricity efficiency in the power block. The total solar-to-electricity efficiency is derived by multiplying the solar-to-heat efficiency by the heat-to-electricity efficiency.

Table D.2 shows the solar-to-heat efficiencies of the solar field and the solar-electricity efficiencies for the full power plant for both the linear concentrator technologies and power towers. These figures were developed by New Energy Externalities Developments for Sustainability (NEEDS) and derived by considering current technologies and an "optimistic-realistic" assessment for future developments [13].

D.13 Solar Thermal Power Density

Mackay [8] estimates the average power density for solar thermal plants in a desert area at 15 W/m^2. This means a 500-MW plant with sufficient storage to run continuously would need a solar field of 33 km^2.

NEEDS reported a higher power density for supplying Europe from the Sahara Desert. The NEEDS report suggests that to supply Europe with 3,500 TWh/year would require an area of 14,400 km^2 at 100% capacity factor that implies a power density of 28 W/m^2 [13].

It should also be noted that there can be a significant difference in power density between summer and winter. This variation can be as much as 10:1 in some locations and will often be 4:1 [14]. This can have a substantial impact when using solar thermal systems for baseload.

Appendix E
Energy Storage

One solution to accommodating variable energy sources in an electricity network is to provide energy storage. Storage can smooth out variability by storing energy during peak production to be later retrieved during low production to smooth out the peaks and troughs in a process called load levelling. Storage can also provide standing reserve and emergency grid restart[7] services to the network, as well as frequency regulation. This appendix provides more technical information about the various electricity storage options available.

E.1 Storage Properties

Most electrical energy resources are either stored chemical, mechanical, gravitational, or fission energy (all known as potential energy) or thermal energy (known as kinetic energy). Fossil fuels are stored chemical energy. Hydropower relies on stored gravitational energy in water at high elevation. Geothermal relies on stored thermal energy (heat) from the earth and nuclear power relies on fission energy from uranium and thorium. All non-variable electrical energy sources rely on these energy rich natural resources.

Table E.1 shows typical properties for various natural resources and storage media. Different specific media, even within the same type, vary in their properties. For example, black coal from one mine might have a different energy density to coal from another mine, and NG might be compressed to increase the energy density.

Those media with a high energy density are able to store large quantities of energy in a smaller volume. Energy density in the table is expressed in megajoules

[7] Emergency restart means providing the initial start-up energy needed for a power station after a grid shutdown.

M. Nicholson, *The Power Makers' Challenge*, Green Energy and Technology,
DOI: 10.1007/978-1-4471-2813-7, © Springer-Verlag London 2012

Table E.1 Properties of energy storage media

Medium	Energy density MJ/m^3	Electrical energy density kWh/m^3	Conversion efficiency (%)	Storage period hours	Comments
Natural resources					
Natural uranium (fast)	150,000,000,000	12,500,000,000	30	–	Fast reactors
Natural uranium LWR	950,000,000	80,000,000	30	–	Thermal
Black coal	24,000	2,300	35	–	–
Brown coal	15,000	1,000	25	–	–
Dry wood	10,000	970	35	–	Biomass
NG	38	5	45	–	CCGT
Man-made storage					
Hydrogen compressed	3,300	600	65	2–24	350 bar
Li-ion battery	1,200	320	95	0.1–12	Bulk and EV
NaS battery	1,100	260	85	7–11	Bulk
Molten salt	900	90	35	0.5–8	CSP
Nickel–cadmium batt.	350	80	80	0.01–12	–
Lead-acid battery	250	60	95	0.01–10	–
Vanadium-redox batt.	120	25	80	1–10	Flow battery
Flywheel	80	20	90	0.001–0.8	–
Compressed air	25	5	70	2–14	CAES (US)
Water (pumped-hydro)	1	0.25	90	12–100	100 m head

From various sources (2010)

(MJ) per cubic meter (m^3). There is always some conversion loss in converting the energy stored in the medium into electrical energy when needed so the table shows the typical conversion efficiency and the electrical energy recovered per m^3.[8]

The reason that natural resources like coal and uranium have been so successful at generating electricity is largely because of their high energy density—particularly uranium.

E.2 Electricity Storage

To smooth out the peaks and troughs from variable energy sources we need to store large quantities of energy for lengthy periods. With a high proportion of wind power for example, the storage requirement could be several days.

[8] 1 kWh = 3.6 MJ. 1,000 MJ/m^3 at 80% conversion efficiency = 1,000 × 0.8/3.6 = 222 kWh/m^3.

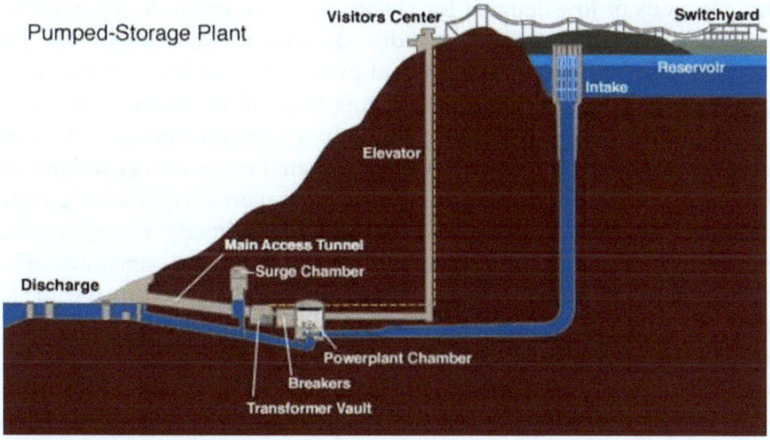

Fig. E.1 Pumped storage. From Tennessee Valley Authority

The effectiveness of an energy storage system depends on a number of factors. "Standing loss" (loss of energy by leakage from the store) needs to be low. 'Cycle efficiency' (retention during transfer to and from the store) needs to be high. A cycle efficiency of one means all the energy transferred to the store can be recovered. This rarely happens of course. A store with a cycle efficiency of 0.5 means that for each kWh of electricity used to charge the store only 0.5 kWh is recovered. Cycle efficiency is sometimes referred to as the 'round-trip' efficiency. The third effectiveness factor is 'energy density'—the electrical energy per kilogram (kg) or per m^3 of storage media that can be recovered from the store. Table E.1 shows typical electrical energy densities for several storage media. A low density figure means a large storage area.

Storage period depends the size of the storage, the storage charging time and the energy demand on the storage system. The storage periods shown in Table E.1 are typical figures for existing installed storage systems.

The rest of this appendix will look at some of the more commonly used bulk storage systems used for stationary energy applications plus some possible systems for the future.

E.3 Pumped Hydro Storage

Pumped Hydro Storage (PHS) relies on the gravitational energy stored in water in a high reservoir that is above a lower reservoir—usually at least 100 m below (often called the 'head'). When this water is released through a pipe or tunnel it flows down to the lower reservoir through reversible water turbine/generators to convert the gravitational energy into electrical energy. The discharged water is then collected in the lower reservoir (See Fig. E.1).

During periods of low demand for electricity in the network, the generators are used as electric motors (see Appendix B—Power Generation and Network Control) that drive the water turbines and push the water back up the tunnel into the upper reservoir. It can take several hours to refill the upper reservoir.

The amount of energy that can be stored in a pumped storage system depends on the amount of water stored and the head. Given the low energy density of water (see Table E.1) we need very large reservoirs to generate significant quantities of electrical energy. The volume of water needed is directly related to the head between the upper reservoir and the turbines as well as the generator efficiency. From Table E.1 the electrical energy density of water with 100 m head is 0.25 kWh/m^3 with a turbine/generator efficiency of 90%. To store 100 GWh would require both reservoirs to be capable of holding 400 million m^3 of water (400 GL). If the reservoirs had an average depth of 20 m the area would need to be 20 km^2 for each reservoir. If the head was doubled to 200 m then the volume needed would be halved to 200 GL.

The full round-trip efficiency for PHS, including both pumping and generating, is around 70–80% so the cycle efficiency is 0.7–0.8. For every MWh used by the pumps to return the water to the upper reservoir we can expect 0.7–0.8 MWh out of the generators when the water is released.

E.4 Compressed Air Energy Storage

The Compressed Air Energy Storage (CAES) process involves storing electrical energy in the form of mechanical energy in compressed air. CAES uses an electric compressor to compress air into an underground storage area using surplus low-cost electricity (See Fig. E.2). When the electricity is needed from the store the compressed air can be released, mixed with NG, heated (by combustion) and used to drive a gas turbine.

The amount of energy that can be stored in a CAES system depends on the volume and pressure of the stored air. Using the electrical energy density of compressed air shown in Table E.1, to store 100 GWh of electricity would require a storage volume of about 20 million m^3. The existing CAES system in the US (see Sect. 10.3) has a storage volume of around 0.5 million m^3.

When generating electricity, a CAES system using NG in the turbine gets about 40% of the energy from the compressed air and 60% from the gas. The cycle efficiency for the compressed air cycle (excluding the gas cycle) is 0.5–0.6. For every MWh of off-peak electricity used to drive the compressor, we can store 0.5–0.6 MWh of electricity.

The reason the current CAES systems use NG is to heat the air before it is used to drive the turbine. The process of compressing the air into storage already heats the air but the heat is removed before storage so that the stored temperature is around 35°C. Currently this heat is discarded. Separately storing and reusing the removed heat in a storage tank would improve the round-trip efficiency of the

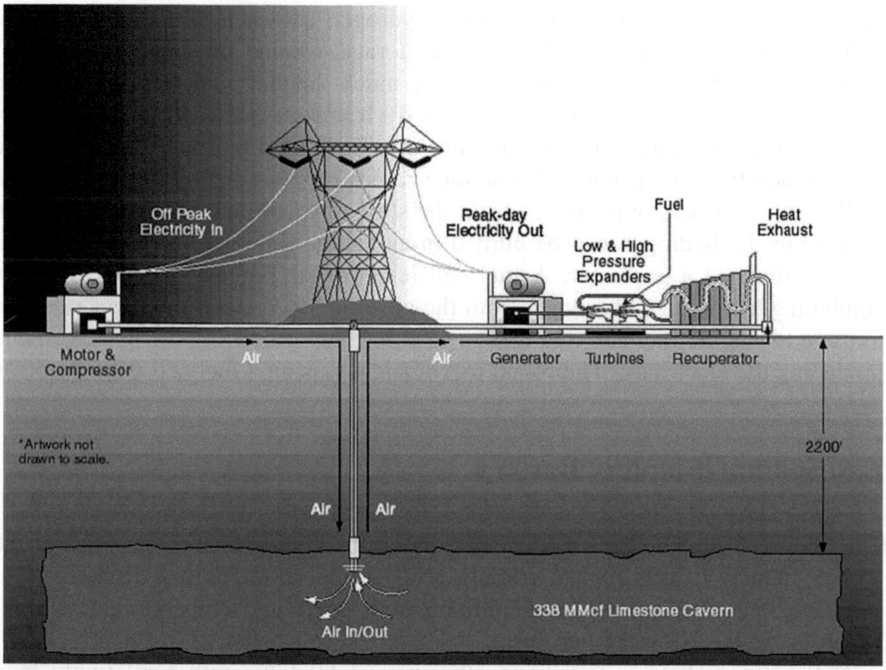

Fig. E.2 Compressed air energy storage. From US Department of Energy

CAES system. The released compressed air could be made hot enough to drive the turbine without using NG.

Such a process called advanced adiabatic CAES (AA-CAES) has been under research for several years. The advantages of AA-CAES are the saving in gas costs as well as greenhouse gas emissions created by burning the gas. It can make CAES a low-emission electricity storage system as long as the electricity used for the compression process was also low-emission. AA-CAES is still in the research phase of development.

E.5 Hydrogen Energy Storage

Hydrogen energy storage uses electrolysis to convert surplus low-cost electricity into hydrogen. Passing an electrical current through water in an electrolyzer releases hydrogen and oxygen. The oxygen can be released to the atmosphere and the hydrogen compressed and stored using a hydrogen gas compressor and storage vessel. The energy density of compressed hydrogen is over 100 times higher than compressed air, so less storage space is needed compared to CAES.

As with compressed air, the amount of energy that can be stored in a hydrogen storage vessel depends on the volume and pressure of the stored hydrogen. Using the electrical energy density of compressed hydrogen shown in Table E.1, to store 100 GWh of electricity would require a storage volume of about 170,000 m^3. In practice hydrogen would be stored in much smaller volumes close to the generator. A 200-MW wind farm might use 10 h storage or 2 GWh at full capacity with a hydrogen storage volume of 3,400 m^3. This could be six spherical storage vessels each 10 m in diameter. The storage could also be in an underground cavern.

When the electricity is needed from the stored hydrogen, either a fuel cell can be used or the hydrogen can be burned in an engine (internal combustion or gas turbine) to drive a generator. A fuel cell reverses the process of electrolysis by combining hydrogen and oxygen from the air to produce electricity. It is possible to use a reversible fuel cell that can be used for both producing the hydrogen and producing the electricity. The round trip efficiency using a fuel cell is under 50%.

E.6 Sodium–Sulphur Battery

The Sodium–Sulphur (NaS) battery has one of the highest energy densities of the battery family. It uses molten sodium (Na) as the anode during discharge and sulphur (S) as the cathode. It requires a high operating temperature (>300°C) which keeps the Na in molten form.

NaS batteries have a round-trip efficiency of 80–90% with a relatively long battery life (10–15 years). They are used for smaller bulk storage needs, up to a few hundred MWh. Based on Table E.1, a 100-MWh system would have a battery volume of around 400 m^3.

E.7 Lithium-Ion Battery

The Lithium-ion (Li-ion) battery has an energy density similar to NaS. It typically uses graphite (carbon) as the anode during discharge and a lithium (Li) compound as the cathode. The compounds used vary with battery type and function. Li-ion batteries can be used for both small bulk storage of 10–50 MWh and in battery-electric vehicles.

E.8 Flow Batteries

A flow battery uses chemical solutions which flow through a power cell where the chemical energy in the solutions is converted to electrical energy. The chemical solutions are stored in large tanks outside the cell and pumped into and out of the cell during operation (See Fig. E.3).

Several different chemical combinations have been used successfully for flow batteries including zinc-bromide, sodium-bromide, and vanadium.

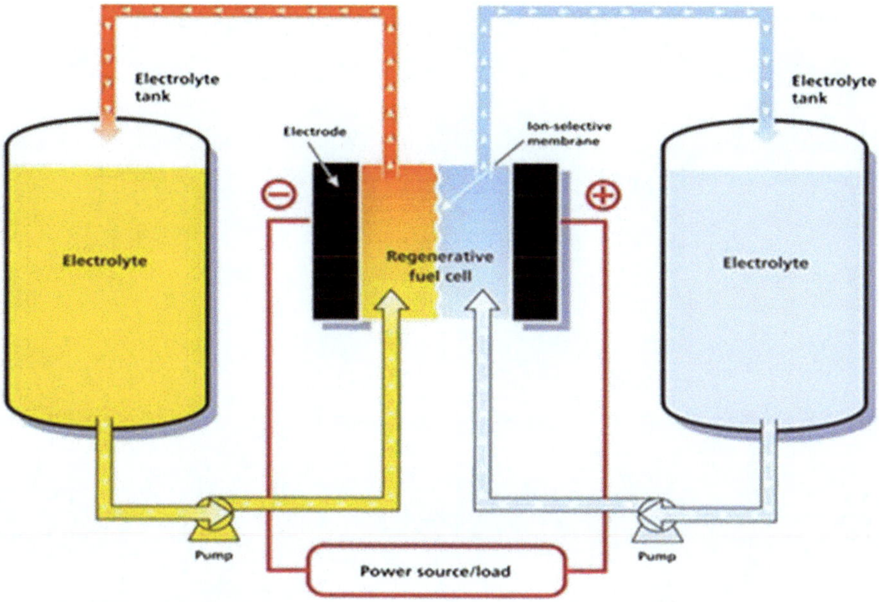

Fig. E.3 Flow battery. From wikis.lawrence.edu

The big advantage of flow batteries is that they can be scaled up or down by using larger or smaller tanks to store the chemicals. They can also be fully charged and discharged without problems unlike conventional batteries. The disadvantage is the relatively low energy–density compared to storing hydrogen as can be seen from the vanadium-redox example shown in Table E.1. Flow battery storage needs about 25 times the storage volume of equivalent hydrogen storage.

Appendix F
Carbon Capture and Storage

In this appendix we cover the carbon capture and storage (CCS) process that involves separating the carbon dioxide (CO_2) from a localised industrial or energy source (such as a power station) and transporting it to a long-term storage location isolated from the atmosphere. With CCS for electricity generation, the localized point sources are large fossil fuel plants such as coal- or gas-fired power stations or biomass-fired power stations. Potential storage locations are geological formations as shown in Fig. F.1.

F.1 Capture Process

The CO_2 capture can either be before the fuel is burned or after. Where the capture is post-combustion (after), the CO_2 is separated from the other flue gases before they are released to the atmosphere. The most common method of separation is called solvent absorption where the CO_2-containing gas is passed through a liquid absorbent that is capable of capturing the CO_2. The sorbent loaded with the captured CO_2 is transported to a different container where it is heated to release the CO_2. This technology is already in use for NG and could be fitted to existing non-CCS power plants. Membranes and solid absorbents can also be used to separate out the CO_2 from the other flue gases.

An alternative post-combustion capture approach called oxy-fuel, burns pulverized coal in nearly pure oxygen rather than air to produce a flue gas of mainly CO_2 and H_2O (water). This simplifies the separation of the CO_2 from the flue gas (described above) but uses additional energy to extract the oxygen (used for the coal burning) from air.

Where the capture is pre-combustion, the coal is gasified to produce syngas which consists of carbon monoxide and hydrogen. The carbon monoxide is reacted with steam in a catalytic reactor to produce CO_2 and more hydrogen. This process is widely applied in the fertiliser and chemical industries. The CO_2 can be separated by a physical or chemical absorption process to produce a hydrogen-rich

DOI: 10.1007/978-1-4471-2813-7, © Springer-Verlag London 2012

fuel which can be used in a gas turbine. This process is called integrated gasification combined cycle (IGCC).

The capture process makes up about 80% of the total cost of CCS. According to the Intergovernmental Panel on Climate Change (IPCC), current technology can capture about 85–95% of the CO_2 in a power plant but would use 10–40% additional energy depending on the fuel and the capture technology used [15].

F.2 Transporting the CO_2 for Storage

Once the CO_2 is captured it must be transported to a suitable storage site. Transport of CO_2 is either by pipeline or shipping.

Pipelines are commonly used to carry gases over long distances. Natural gas (NG) is transported this way and NG pipelines can cover thousands of kilometers. CO_2 is already transported by pipelines to enhanced oil recovery (EOR) projects in several locations. Pipe corrosion needs to be minimized by drying the CO_2 and removing any hydrogen sulphide. Pipelines that run through populated areas also need to be monitored for leaks. The CO_2 is usually compressed to reduce the volume of gas before transporting by pipeline.

Like NG, CO_2 can be liquefied to reduce the volume even further for transportation by ship tankers or road and rail tankers. Road and rail transportation is likely to be too expensive for large-scale projects being twice the cost of pipelines. The IPCC believes that many large fossil fuel plants are within 300 km of potential geological storage sites [15].

F.3 Storage Sites

CO_2 can be stored underground (see Fig. F.1) in depleted oil and gas reservoirs, un-minable coal beds and deep underground porous reservoir rocks saturated with brackish water (saline aquifers). The CO_2 is compressed before storage in these deep geological formations. These formations can be both onshore and offshore under the sea bed.

The technologies used to inject CO_2 into deep geological formations were developed for the oil and gas industry. At depths below 800–1,000 m where the pressure is high, CO_2 has the density of a liquid. This makes the use of underground storage in the pores of sedimentary rocks both more efficient and more secure. The CO_2 is prevented from migrating to the surface by remaining trapped in the pore spaces and by a layer of impermeable rock which prevents fluid flow from the storage area. At shallower depths, such as in coal beds, the CO_2 can be absorbed on the coal surface.

Liquid CO_2 could also be stored at great depth in the oceans where it would stay for centuries. The CO_2 can be transported by pipeline or ship and released on the ocean floor. Below 3,000 m, CO_2 is denser than seawater and would form a CO_2

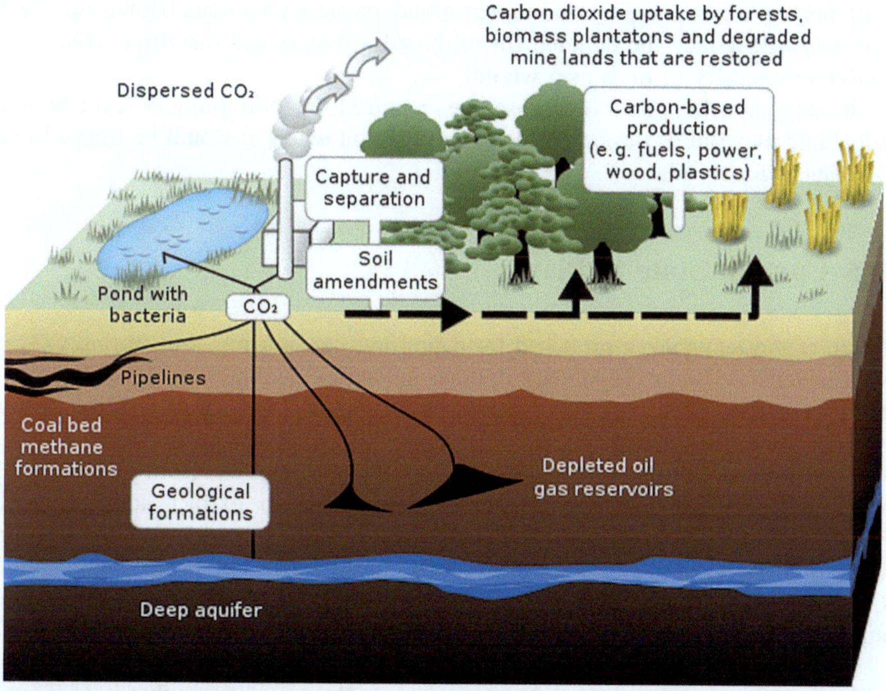

Fig. F.1 Carbon capture and storage. From Wikimedia Commons Hardin L, Payne J modifications made by Arntzen (2009)

lake that would slowly dissolve into the surrounding seawater. The average depth of the oceans is 3,800 m so there are ample potential deep storage sites.

The oceans cover over 70% of the Earth's surface and already absorb CO_2 from the atmosphere but most of this remains in the upper ocean. Deep ocean storage is still in the development stage. There is a concern that significant volumes of CO_2 in the ocean will change the ocean chemistry and significantly impact marine life.

F.4 Bio-CCS

An alternative to storing the CO_2 is to use it in an algae farm (see Fig. F.1) to grow algae that can be converted into oil products such as biodiesel, jet fuel, synthesis gas, and plastics as well as fertilizers and animal feedstock. Some pilot algae plants have already been constructed. It is anticipated that a 5,000 ha algae farm could process half of the CO_2 from a large coal power station [16]. 5,000 ha is about the area of a strip mine that could supply the power station with the coal (see Table 11.3).

Naturally some of these products, such as the biodiesel, will return the CO_2 to the atmosphere when it is burned in an engine so the storage is not permanent and will not be a replacement for underground or ocean storage. However, these biofuels will reduce the demand for oil-based products and thereby decrease the emission intensity of fuels as a whole.

It may also be possible to use the captured CO_2 to produce cement and aggregate material or through mineral carbonation where it would be trapped in a permanent form [17].

F.5 Current State of Play

Much of the technology proposed for capturing, transporting, and storing CO_2 is well understood and used already in other industries. The issue will be whether it can be scaled up to cope with major point source CO_2 producers like power stations, and at what cost.

CO_2 is already stored in underground geological formations from EOR projects where CO_2 is injected into declining older fields to extract additional oil. Thousands of kilometers of CO_2 pipeline already exist to transport the CO_2 to these EOR sites. EOR is mature technology but the scope for further high volume CO_2 storage is limited. However, existing abandoned oil and gas fields are attractive possible sites for future CO_2 storage.

Saline aquifers have been used for storing chemical waste in the past. These saline aquifers are abundant and contain large volumes of water that is of no use for agriculture but could be used for storing CO_2. Commercial scale CO_2 storage in saline aquifers is already done in the Sleipner Project in the North Sea [15].

There is a flurry of activity around the world to build pilot capture and storage facilities that can be scaled up to large power station dimensions. But outside a few EOR sites and the Sleipner Project no large-scale projects have yet been commissioned.

Appendix G
Nuclear Power

In this appendix we discuss how nuclear reactors work including the fission process to generate heat. We cover nuclear fuel processing and cycles and discuss several nuclear technologies both present and future (Fig. G.1).

G.1 Nuclear Fission

Nuclear fission occurs when an atom of what is called fissile material is split apart by a 'slow' neutron. This process, due to small mass differences between the parent and daughter nuclei, releases large amounts of energy in the form of heat and more neutrons. The fissile material most commonly used for nuclear power is a specific isotope of uranium called U-235.

If these generated neutrons are then absorbed by other uranium nuclei, and at least as many neutrons are created as are consumed, a chain reaction begins. In a thermal nuclear reactor, materials called moderators are used to control this chain reaction process. These moderators absorb the energy in impacting neutrons and slow the neutrons down. Slow neutrons are more readily absorbed by U-235 than the original fast ones. A controlled chain reaction takes place when about 40% of the neutrons produced go on to cause subsequent reactions (Fig. G.2).

Most current nuclear power plants use the heat generated in a controlled fission reaction to produce steam. The steam is used to drive conventional steam turbines and generators. We discuss how this is done when discussing reactors below.

G.2 Nuclear Fuel Processing

The most commonly used fuel in nuclear power plants is uranium. Uranium is found naturally as an oxide mineral in rocks. It is quite common and widely distributed throughout the world. It occurs in low concentration and must be mined and

M. Nicholson, *The Power Makers' Challenge*, Green Energy and Technology,
DOI: 10.1007/978-1-4471-2813-7, © Springer-Verlag London 2012

Fig. G.1 Nuclear power plant. From Lange P (2009)

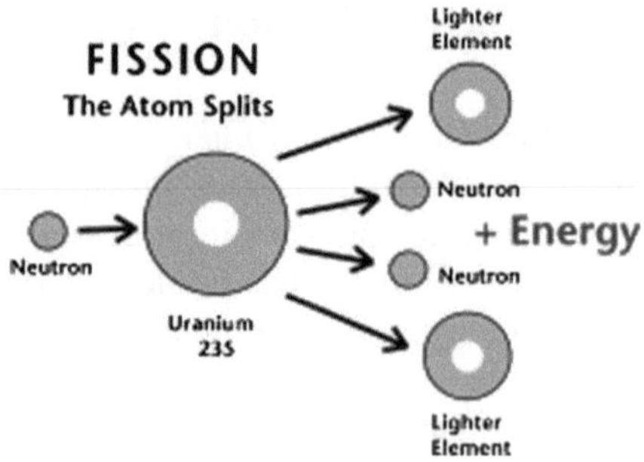

Fig. G.2 Fission energy. From US Energy Information Administration (2008)

extracted by milling and/or chemical treatment processes. About 170 t of uranium is processed in order to run a 1 GW nuclear reactor for a year (see Fig. G.3).

The uranium oxide concentrate (sometimes called yellowcake) is mainly U_3O_8. Naturally occurring uranium is composed of 99.3% non-fissile (also referred to as 'fertile') U-238 with a small amount (0.7%) of fissile U-235. Most nuclear power plants only use the U-235 as a fuel so they use less than 1% of the naturally occurring uranium. To sustain a nuclear chain reaction in most nuclear reactors, the uranium oxide has to be enriched to increase the concentration of U-235 to above 3% by removing much of the U-238. This is then called low-enriched uranium.

Enriched uranium is transformed in a fuel fabrication plant into another oxide of uranium, UO_2. UO_2 is a black powder that is compressed and baked into small ceramic pellets which are then stacked into thin metal tubes to create long fuel rods. These rods are then bundled into fuel assemblies for insertion into the reactor. To control the rate of nuclear reaction in the fission process, control rods are inserted into the fuel rod bundles.

These control rods are made of materials such as silver-indium-cadmium alloys or boron, which absorb neutrons and slow the reaction. To increase the reaction and the amount of heat generated, the control rods are raised out of the fuel rod bundles. If the control rods are fully inserted into the bundles the reaction stops and allows the reactor to be shut down. A reactor can contain several hundred fuel rod assemblies and a 1,000 MW reactor uses about 75 tonnes of low-enriched uranium at any one time.

Thorium is potentially an alternative fuel source for nuclear fission. Thorium (or more specifically the common thorium isotope Th-232) is, like U-238, a fertile material which means it can be converted to fissile material. Thorium cannot be used directly as a substitute for uranium, but it will absorb slow (low energy) neutrons to produce the fissile uranium isotope U-233. Initially thorium needs to be used with a fissile material like U-235 or the plutonium isotope Pu-239 so it can provide a source of U-233 to sustain the process. The interest in using thorium arose partly because it is more abundant in nature than uranium, but also because all of the mined thorium is potentially useable in a reactor, compared with the 0.7% of natural uranium. The increased cost and complexity, and lack of sustained investment by research organisations for the fuel cycle compared to U-235, has held back its development.

G.3 Nuclear Fuel Cycles

Current thermal reactors use either a once-through (open) cycle where the fuel is used once and the spent fuel stored and disposed of, or a closed cycle where the spent fuel is stored and then reprocessed. Most reactors generally use a once-through cycle.

The spent fuel rods from the reactor are replaced with new ones every 12–24 months and about 20 t of fuel is replaced each year. The spent fuel is still highly radioactive and produces a lot of decay heat from the fission products. The spent fuel rods are initially stored in ponds like large swimming pools close to the reactor.

In a closed cycle, nuclear fuel is supplied in the same way as in an open cycle, but when the spent fuel rods are removed from the reactor they are reprocessed and some fissile material is recovered. Spent fuel still contains 95% of its original energy and some of this can be recovered with reprocessing.

Reprocessing separates the radioactive spent fuel into uranium and plutonium for re-use leaving some long lived waste fission products for disposal (see Fig. G.3).

Reprocessing makes it possible to recycle the uranium and plutonium into fresh fuel and reduces the amount of waste. The current standard method for reprocessing is the aqueous method called PUREX (Plutonium–URanium Extraction). A more advanced proliferation-resistant reprocessing method using pyro-metallurgical processing (pyro-processing) is under development, and may be favored for next generation reactors like fast breeder reactors (FBR).

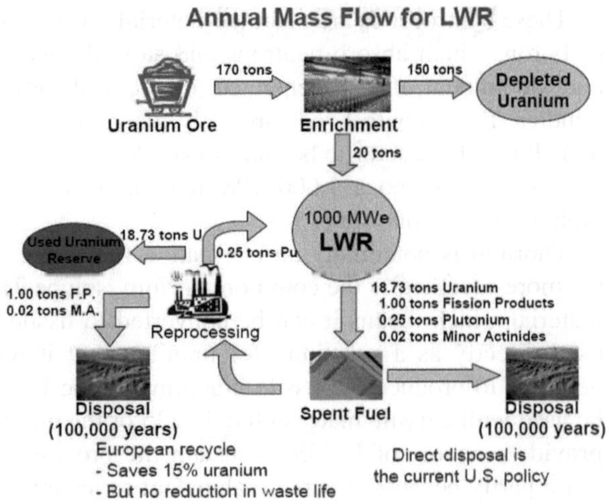

Fig. G.3 Nuclear fuel cycles. From Chang Y (2010)

FBR can use a much greater percentage of the initial fissionable material than thermal reactors, and can actually create more fissile material than they started with, by transmuting fertile isotopes into fissionable forms. With recycling of fissile and fertile materials, breeder reactors can use almost all of the initial fissionable material. Approximately 60 times[9] more energy can be extracted from uranium by the FBR cycle than from an open cycle thermal reactor [3].

Breeder reactors have been under development since the 1950s and can be designed to use thorium as well as uranium. The development has been a low priority due to the higher cost of breeder reactors and an abundance of cheap uranium. We discuss FBR in more detail below.

G.4 Nuclear Reactor Technology

Nuclear reactors for power generation were first developed in the 1950s. Since then there have been three generations of reactor technology and the Generation IV International Forum is overseeing the development of the next generation.

All present day nuclear power plants use nuclear fission and most use uranium as a fuel source. The predominant designs are thermal plants, with only a few fast reactors currently delivering electricity to commercial grids (Fig. G.4).

[9] 60 times is possibly very conservative. It could actually be 160 times, see Sect. 18.4.

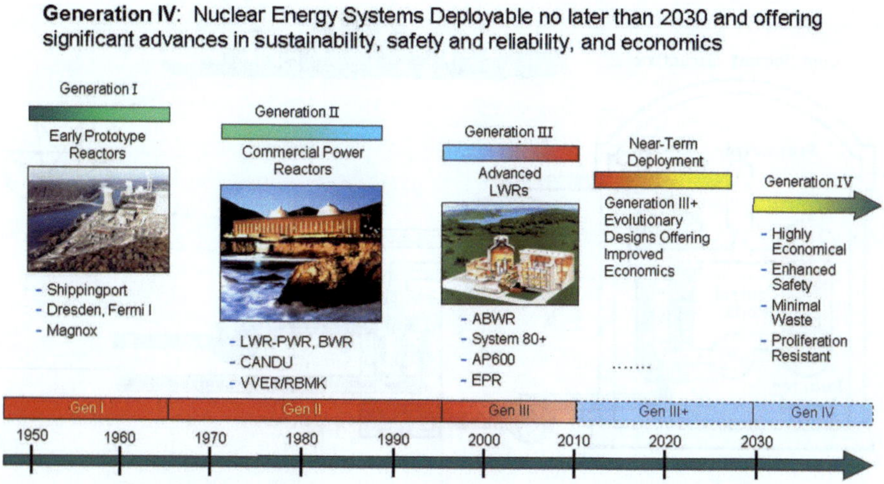

Fig. G.4 Generation IV reactors. From US Department of Energy (2010)

G.5 Thermal Reactors

Nuclear power plants are used to harness and control the heat energy released through nuclear fission All plants operate on the same principle, but different designs are currently in use throughout the world. More than half of the power reactors in use today are pressurised water reactors (PWR). The next most popular design is boiling water reactors (BWR). PWR and BWR are known as light water reactors (LWR). Other thermal reactors in use are pressurised heavy water reactors (PHWR). The three types vary in operating conditions and fuel mixes, but the basic principles are similar. The reactors of these types built between the 1960–1990s are often referred to as Generation II reactors.

PWR use ordinary water as both coolant and moderator in the reactor core. As the water in the primary coolant loop passes through the reactor core, it is heated to over 300°C by the fission process in the reactor vessel (see Fig. G.5). The water is held at high pressures to prevent boiling. It transfers thermal energy to a secondary loop in the steam generator producing steam to drive the steam turbine. The water in the two loops does not mix. This avoids the radioactive water in the primary loop getting into the steam turbine.

In a BWR, water is constantly fed into the bottom of the primary vessel and boils in the upper part of the reactor core (see Fig. G.6) the steam generated is passed directly to the turbine in a single primary loop.

The BWR design has several advantages over the PWR. It does not require a separate steam generator and operates at lower pressure. This reduces the reactor vessel wall thickness and material costs. The BWR also has greater thermal efficiency than a PWR as there is no heat exchanger/steam generator.

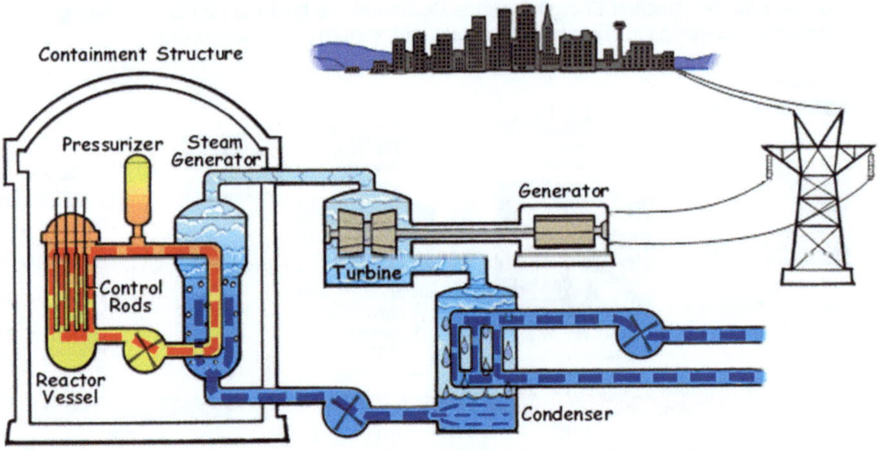

Fig. G.5 Pressurised water reactor (*PWR*). From US Nuclear Regulatory Commission (*NRC*) (2009)

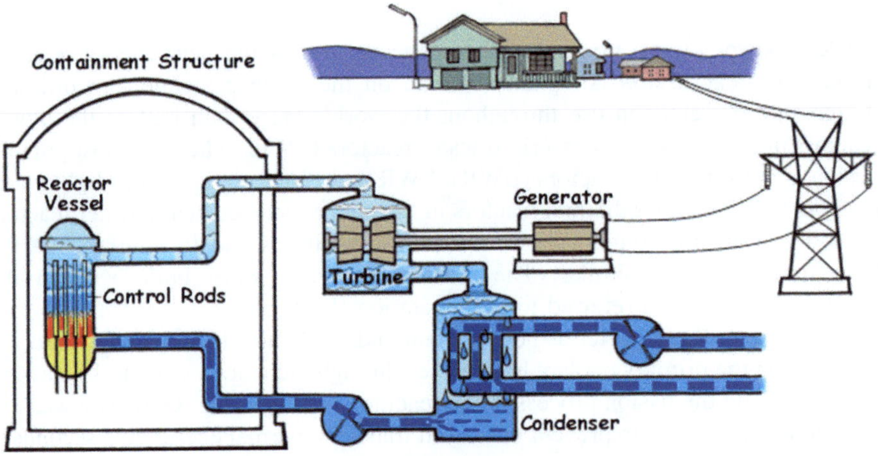

Fig. G.6 Boiling water reactor (*BWR*). From US Nuclear Regulatory Commission (*NRC*) (2009)

However, the BWR primary loop includes the turbines which become radioactive over the lifetime of the reactor. This complicates plant maintenance and increases the costs of decommissioning. The increased cost of BWR maintenance tends to balance the savings from the simpler design and greater efficiency.

The PHWR/CANDU[10] design is similar to the PWR but uses heavy water (deuterium oxide D_2O) rather than light (distilled) water for both coolant and

[10] CANDU stands for CANada Deuterium Uranium.

moderator. Heavy water does not absorb neutrons as readily as light water and this frees up more neutrons to react with fissile isotopes. As a result, natural uranium dioxide containing 0.7% U-235, can be used as a fuel without enrichment. The cost saving of using non-enriched natural uranium is partially offset by the increased cost of the heavy water and the faster consumption of the non-enriched fuel.

The CANDU fuel bundles and coolant are contained in hundreds of pressure tubes in a large tank of heavy water moderator. This negates the need for the single thick-walled steel pressure vessel of a PWR. An advantage of this pressure-tube design is that it can be re-fuelled while operating at full power. PWRs and BWRs need to shut down for 30–60 days every 18–24 months to replace around one third of the fuel load. This on-load refuelling improves the CANDU capacity factor.

Advanced CANDU designs (ACR) have been developed that use light water for the coolant keeping heavy water as the moderator. This reduces the use and cost of the heavy water, but none have yet been used commercially.

Generation III reactors comprise several advanced light water reactor designs.

The Advanced BWR (ABWR) was developed in the 1990s. Improvements to BWR design include better economics, passive (replacing some active) safety features, better fuel utilisation and reduced waste. Passive systems use gravity and natural circulation avoiding the need for pumps, fans, or other rotating machines that require an external power source that may fail.

The European or evolutionary pressurised reactor (EPR) is an advance PWR design. As with the ABWR the focus has been on increased safety while providing enhanced economics. The first two units are under construction in Finland and France. Another advanced PWR design is the Westinghouse AP1000. It is claimed to be the safest and most economical nuclear power plant available today. The plant design is simplified to shorten the construction time and to make it easier and less expensive to build, operate, and maintain. It also includes passive safety features. The first plants are currently being constructed in China.

Two future generation thermal reactor systems under development are gas-cooled rather than water-cooled and use graphite as a moderator. The pebble bed modular reactor (PBMR) and the gas turbine modular helium reactor (GT-MHR) use helium as the coolant and operate at much higher temperatures than current light water reactors and have greater thermal efficiencies and higher fuel burn-up rates. The helium is used to drive a gas turbine rather than generate steam for a steam turbine.

Unlike light water reactors, where the uranium oxide fuel is in the form of pellets enclosed in a metal tube (see Sect. G.2 above), the fuel for these gas-cooled reactors is in the form of tiny (sub-millimeter) spheres and can be uranium, thorium or plutonium. The spheres have a core of enriched fuel coated with layers of ceramic. Thousands of the fuel particles are pressed together and coated with an external layer of graphite moderator. In PBMRs, the pressings are in the shape of tennis-ball size spheres called "pebbles". In GT-MHRs the fuel is in small cylindrical rods.

Both these reactors can be on-load refuelled and the fuel is claimed to have excellent proliferation resistance. The high operating temperature allows a deep burn of the fuel resulting in less long-lived waste which is also easier to store and transport. The downside is that the fuel is difficult to recycle.

G.6 Fast Reactors

Fast reactors are considered Generation IV technology and use high energy neutrons directly to sustain a fission chain reaction. Unlike thermal reactors, fast reactors do not have moderators. As well as burning the fissile fuel through fast fission, fast reactors can activate fertile material like thorium (Th-232) or U-238 to create additional fissile material. They can also "burn" long-lived radioactive waste in spent fuel from conventional thermal reactors. The remaining waste is about one-fiftieth the Vol. of current spent fuel and decays to be less radioactive than natural uranium ore within a few hundred years which significant reduced the nuclear waste problem.

Fast reactor fuel needs to contain at least 20% fissile material such as U-235 or Pu-239. This is much higher than the fissile percentage of the low-enriched uranium needed for thermal reactors (3–5% U-235). The other 80% of the fuel can be fertile material like U-238 which can come from natural uranium or reprocessed uranium from traditional light-water thermal reactors.

Fast reactors use a much higher proportion of the mined uranium than traditional thermal reactors because they use the abundant fertile U-238 that is often discarded in open-cycle thermal reactors (see Sect. G.3 above). Fast reactors transform this non-fissile U-238 into fissile Pu-239.

This allows them to generate much more energy from the uranium. In the 1950s, there were concerns about the long-term supply of mined uranium so FBR were developed to improve the long-term viability of nuclear power. A reactor is called a "breeder" when it produces more fissile material than it consumes. A typical fast reactor could contain a core comprising several thousand stainless steel tubes containing 20% Pu-239 mixed oxide fuel. This core is surrounded by a blanket of rods containing uranium oxide or thorium. By changing the material used in the blankets, a fast reactor can be operated over a spectrum from being a pure breeder reactor to a self-sufficient converter. In breeder mode (using U-238 blankets) it will produce more fissile material than it consumes. This is useful for providing a surplus of fissile material for starting up other reactors. Using steel reflectors instead of U-238 blankets, the reactor operates in pure burner mode and heats the coolant to generate steam in a heat exchanger that can be inside or outside the reactor tank (see Fig. G.7). The steam can then power a conventional turbine/generator combination.

Current fast reactors use liquid metal as a coolant. Early reactors used mercury or lead. Future fast reactors will probably use molten Na—sometimes referred to as Sodium-cooled Fast Reactors (SFR).

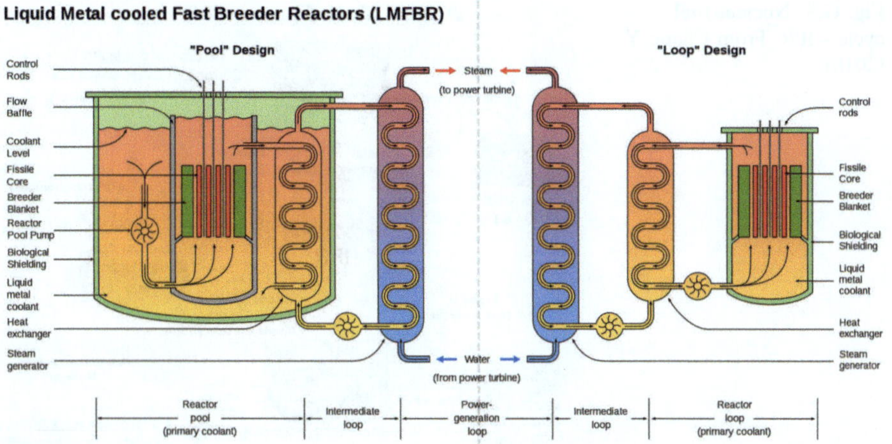

Fig. G.7 Liquid metal cooled fast breeder reactor. From Wikimedia Commons Graevemoore (2008)

G.7 Integral Fast Reactor

The integral fast reactor (IFR) is a form of liquid metal fast breeder reactor shown above using molten Na as a coolant and combined with a pyro-processing recycling facility. This is possible because the fuel in the IFR is metal, rather than oxide. This not only permits electrochemical recycling, but also simplifies the fuel fabrication process (via injection casting rather than fine-scale milling), and provides exception passive safety benefits.

Originally developed by scientists at Argonne National Laboratory in the US between 1984 and 1994, the IFR promised safe nuclear power, not limited by fuel supplies, with a waste product reduced in both quantity and radioactive lifetime.

The original project was stopped in 1994, believed to be for political reasons, but with the growing concerns about greenhouse gas emissions and the long-term availability of mined nuclear feedstock, there has been a push to have the project restarted. The science is well established but the technology still needs to be fully developed. Both India and China are in the process of developing liquid metal FBR (Fig. G.8).

G.8 Nuclear Fusion

In contrast to the fission process described above, fusion is a process in which light elements, such as hydrogen, collide and combine or fuse to form slightly heavier elements such as helium. This process releases large amounts of energy. Nuclear fusion occurs naturally in the centre of stars like our sun.

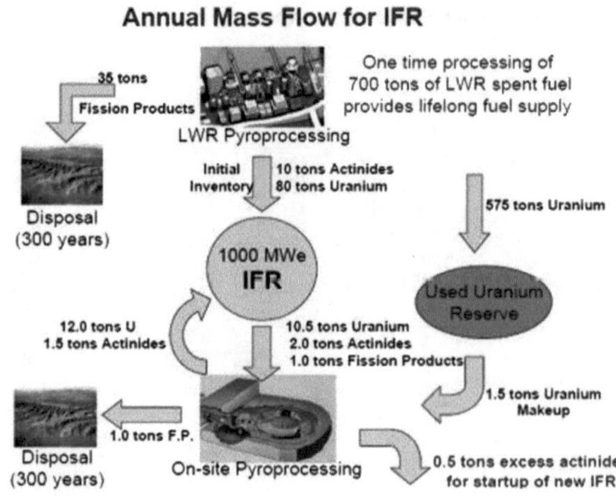

Fig. G.8 Nuclear fuel cycle—IFR. From Chang Y (2010)

Nuclear fusion for electricity generation offers two major potential benefits over fission thermal reactors. First, the reactor fuel (deuterium—a stable isotope of hydrogen) can be obtained from the ocean providing a virtually unlimited fuel source. Second, fusion would produce no long-lived radioactive waste products. Note these are both also true for fast fission reactors.

It also offers some significant challenges. The conditions require for fusion are very high temperatures in excess of 150 million °C (10 times hotter than the sun's core) and high pressures using intense magnetic fields. The challenge is to produce these conditions in a controlled environment at a reasonable cost. The most efficient fusion reaction for a power reactor is between deuterium (D) and tritium (T). The D-T fusion reaction produces the highest energy gain at the "lowest" temperatures. D-T fusion produces one Helium nucleus, one neutron and energy.

The world's first controlled release of fusion power was achieved in the JET Tokomak in the UK in 1991.

The most significant fusion research activity currently is the international thermonuclear experimental reactor (ITER). ITER is an international collaboration between the European Union, Japan, the Russian Federation, the United States, China, the Republic of Korea, and India. The ITER project is located in France and is estimated to cost around 10 billion Euros over 10 years [18]. ITER aims to develop the technologies essential to proceed towards a functioning fusion reactor, including components capable of withstanding high neutron and heat flux environments.

The ITER site preparation was started in January 2007 and fusion experiments are expected to begin in 2018. It is planned that fusion power will be put into the grid as early as 2040. It is still unclear what contribution fusion will make to greenhouse gas mitigation by 2050, if any.

Appendix H
Electricity Costs and Markets

In this appendix we will take a more technical look at how electricity generating costs are calculated. There are three distinct costs that are referenced: the "levelized cost of electricity" (LCOE), the "short-run marginal cost" (SRMC) and the "long-run marginal cost" (LRMC). We will also look at electricity markets and how they are impacted by generating costs.

H.1 Levelized Cost of Electricity

The LCOE is the average cost per MWh delivered by a power station over its lifetime. It can be used for several purposes, including:

- Estimating the costs of producing electricity for a given technology.
- Identifying the least-cost option among alternative generation investments.
- Evaluating the impact of market changes, such as changes in fuel prices, on generation costs.
- Assessing the impacts on unit costs of changes in key assumptions, including key policy parameters such as carbon prices and interest rates.

To calculate the cost, we need to consider the following factors:

- The capital costs of building the power plant (sometimes referred to as the 'overnight' cost—what it would cost if it were miraculously built overnight).
- The cost of operating and maintaining the plant.
- The fuel price (which would be zero for solar and wind).
- The carbon price.
- The construction time before electricity is delivered and revenue starts to flow.
- The discount rate (see below).
- The capacity factor of the plant.
- The lifespan of the plant.
- The decommissioning cost at the end of the plant's life.

M. Nicholson, *The Power Makers' Challenge*, Green Energy and Technology,
DOI: 10.1007/978-1-4471-2813-7, © Springer-Verlag London 2012

For those familiar with discounted cash flow, the levelized cost methodology discounts the time series of expenditures to their present values in a specified base year by applying a discount rate. The discount rate that is considered appropriate may differ from country to country, and from utility to utility and also differs between technologies. Applying a discount rate takes into account the time value of money—a dollar earned or spent in the past or in the future does not have the same value as a dollar (in real terms) earned or spent today. The discount rate is typically related to the rate of return that could be earned on comparable investments and usually reflects the financial risks of the particular technology. Typical discount rates applied for LCOE calculations are between 5 and 10%.

The LCOE is the net present value (NPV) of the total cost of building and running the plant divided by the NPV of the total electricity generated over the plant's life. For the mathematically inclined, the formula applied to calculate the LCOE is as follows:

$$\text{LCOE} = \Sigma_t\left[(\text{It} + \text{Mt} + \text{Ft} + \text{Ct} + \text{Dt}) * (1+r)^{-t}\right] / \Sigma_t\left[\text{Et} * (1+r)^{-t}\right]$$

Where:

It Investment cost in year t
Mt Operations and maintenance costs in year t
Ft Fuel cost in year t
Ct Carbon cost in year t
Dt Decommissioning cost in year t
Et Electricity generation in year t
r Discount rate.

LCOE values are significantly impacted by discount rates. A lower discount rate produces a lower LCOE, and this has the greatest impact on plants that have high capital costs but low fuel costs and long life spans such as nuclear.

H.2 Short-Run Marginal Cost

The marginal cost of producing an item is the additional cost of producing one more unit of that item. For electricity, the marginal cost would be the cost of generating one more unit of electricity—say one more MWh.

The SRMC of electricity assumes the additional cost of generating that additional MWh does not require any additional generating capacity. It would just include the additional fuel cost and any additional operations and maintenance costs. Clearly the SRMC will vary from generator to generator but it can also vary with load. The thermal efficiency of a fossil-fuel plant is reduced when it is part-loaded (see Appendix B.2) so it will use more fuel and the SRMC will be higher. For a CCGT plant the efficiency can fall from 55% at full load to 40% at half load.

Similarly, a plant in standing reserve that needs to be cold started will use more fuel to get going and will have a higher SRMC than a spinning reserve plant.

Typically, plants with a low SRMC (like coal and nuclear plants) service baseload. Plants with medium SRMC service intermediate load and high SRMC plants (like OCGT) service peak load.

H.3 Long-Run Marginal Cost

The LRMC is the additional cost of generating one unit over an extended period of time and assumes all costs are included, including additional capital costs. So the additional cost of generating one MWh could be spread across the life of the plant. Clearly for a newly built plant, if these costs are discounted then the LRMC will be the same as the LCOE.

H.4 Brief Description of Electricity Markets

Electricity is a commodity that can be bought and sold, like any other commodity. In an electricity market, generators can offer to sell electricity on the market and users can bid to buy it—much like with any commodity market. Markets vary between regions but generally they all use a wholesale pool where bids and offers are cleared and settled by the market operator. In addition to the wholesale market, buyers and sellers can enter into forward contract or purchase exchange traded options. We will discuss each of these here.

H.5 Wholesale Markets

Generators sell their electricity into the market. The buying wholesale customers are generally retailers that on-sell the electricity to business or household end-users. Some large end-users also buy directly on the wholesale market. The market operator aggregates and dispatches the supply offers from the generators to meet the buyer demand bids. This matching of supply to demand is done in real time and the market/system operator issues instructions to each generator to produce the required quantity of electricity. In this process, the system operator is responsible for maintaining network security and delivering electricity at the lowest available cost.

The market operator issues demand and supply forecasts to assist buyers and sellers in making bids and offers. The generators make offers ahead of each trading day to produce a certain quantity of electricity at various prices for each dispatch period in that day. These dispatch periods could be from 5 to 30 min duration. The prices offered are typically based on the SRMC of the generator. The market operator takes all the offers in each dispatch period and dispatches the generators in order of offer price (called the merit order) with the cheapest generators dispatched first until sufficient electricity is dispatched to satisfy the demand in that period. This ensures that demand is met at the lowest possible cost. Sometimes the dispatch order may be modified based on factors such as generator ramp rates and transmission congestion. The dispatch price for the period is the offer price of the highest priced generator dispatched. This provides an incentive to the generating suppliers to have the most efficient, lowest cost generators to earn the greatest profit.

Some generators (like coal and nuclear) need to run continuously so they may make offers in some periods at lower prices than the SRMC (which could even be a negative price in some markets) to ensure they get dispatched in those periods. Gas peaking generators that do not need to run continuously, and have higher SRMCs, will only make offers in periods when they know the dispatch price is likely to be high.

The wholesale spot price paid to the generators is often set over a longer trading interval than the dispatch period and could be 30 min or 1 h. The spot price is determined by taking an average of the dispatch period prices during that trading interval. This spot price is also the price the buyers have to pay for the electricity consumed in that trading interval.

Given that the dispatch prices are based on the highest offers from the accepted generators in the dispatch period, the spot price can (and does) vary substantially throughout the day. For example, during peak times when available generation capacity may be in short supply, the dispatch prices and hence spot prices could rise considerably. During night-time, when there is potentially plenty of available generating capacity with low SRMCs, then the spot prices are usually at their lowest.

H.6 Recovering Generating Costs

At the start of this appendix we discussed the LCOE for a power station. This is the average cost per MWh delivered by a power station over its lifetime. The power station owner needs to recover this cost plus a profit margin so the owner needs to receive sufficient income from the power station to do this.

As we discussed above, the spot price actually paid by the wholesale market is not directly related to the LCOE. In fact the spot price really reflects the dynamic demand and supply of electricity in the market. The power-station owner needs a way of locking in his return and not just rely on the ups and downs of the wholesale spot price. This can be done by entering into forward contracts.

H.7 Forward Contracts

Forward contracts allow the generators and the electricity buyers to buy or sell a given quantity of electricity at a fixed price over a specified time in the future. The contracts usually relate to a nominated time of day.

For example, a retailer might contract to buy 20 MW of peak load in a 3-month period at $50 per MWh. For any peak load electricity purchased in that period, if the peak load spot price is above $50 the seller of the forward contract pays the difference to the retailer. If the spot price is below $50 then the retailer pays the difference to the seller. In effect both parties are guaranteed to get a price of $50 for any electricity supplied or bought under this contract.

Forward contracts could be directly between a retailer and a generator, as these parties face opposing risks in the wholesale market. They can also be arranged by brokers between their generator and retailer clients. Participants can also purchase future contract on a registered trading exchange to achieve the same end result.

H.8 Options

Whereas future or forward contracts give price certainty, it does lock the parties into defined contract terms for volume and prices when the market prices might move against the agreed contract price.

An alternative is to buy an exchange traded option. An option gives the holder the right to enter into a contract on agreed terms at some date in the future without the obligation of completing that contract. A ‘call’ option gives the holder the right to buy a specific amount of electricity in the future at a predetermined price. Conversely a “put” option gives the holder the right to sell a specific amount of electricity.

A retailer could buy a call option to protect against rising wholesale electricity prices but could then abandon that option if the prices do not rise as expected and buy at the lower spot price instead. Likewise, a generator could buy a put option to protect against falling prices.

There are several different types of options to protect against different possible eventualities. For example, a retailer may be happy to buy at the spot price but just wants to protect the upper price by capping the maximum price they will pay. Likewise, a generator may want to guarantee the minimum price they will receive by setting a floor on the price.

H.9 Generator Profits

Generator operators make a profit in periods in which the price paid for the electricity exceeds their marginal cost. Of course they must do this often enough to also recover their investment costs of building the power plant.

Baseload plants like coal, nuclear, and hydro tend to be expensive to build but less expensive to operate. This means they have low marginal costs and can compete in every period of the day so they generally run all the time except when shut for maintenance or, in the case of hydro, there is shortage of water resources. This allows these plants to recover the high cost of building the plant.

During peak load periods, electricity demand is high and there is a need for flexible generators that can ramp up and down quickly to match the varying load (see Chap. 19). These generators, like open-cycle gas turbines, are often started from cold and use more fuel to start and operate. As a result they have higher marginal costs but are generally much cheaper to build. Because more generation capacity is needed during peak demand, more generators need to compete in the market and the electricity price tends to surge. Remember, the electricity price is set by the highest-price generator dispatched. Plants with a high marginal cost can only make a profit during peak times.

H.10 Impact of Wind on Pricing

Wind farms have a low marginal cost because there is no fuel cost. In networks that do not schedule wind because the wind penetration in the network is low (<10%), wind farms can sell electricity into the grid at anytime depending on wind availability and transmission and load constraints. They are"'price-takers" which means the electricity price is set by the scheduled generators bidding into the market.

Where wind is scheduled, then wind farms will make offers into the market like other generators. Because wind farms have low marginal cost, they will be dispatched ahead of other generators, providing the system operator is satisfied they can deliver the quantity of electricity offered. This requires sophisticated wind forecasting systems as discussed in Appendix C.8. This could have the impact of lowering electricity prices during periods of high wind availability [19]. The system operator will probably limit the amount of electricity they are allowed to generate in a dispatch period to the offer based on guidance from the forecasting system. If the wind strength proves to be greater than forecast then the wind farm will be required to curtail the output to their offer. If the wind strength is less than forecast, the system operator will need to draw on available reserves to keep the network in balance.

References

1. Phuc N (2009) Chapter 4: introduction to rotating machines. http://cnx.org/content/m28323/1.1/. Accessed 28 July 2011
2. Freris L, Infield D (2008) Renewable energy in power systems. Wiley
3. MacKay D (2008) Sustainable energy—without the hot air. UIT Cambridge
4. IEA Wind (2007) Task 25 design and operation of power systems with large amounts of wind power. http://ieawind.org/AnnexXXV/Publications/W82.pdf. Accessed 19 July 2011
5. OECD/IEA (2005) Projected costs of generating electricity—2005 update, OECD. http://www.oecdbookshop.org/oecd/display.asp?sf1=identifiers&st1=662005011P1. Accessed 28 July 2011

6. Strbac G (2002) ILEX energy consulting—quantifying the systems costs of additional renewables. http://www.berr.gov.uk/files/file21352.pdf. Accessed 28 July 2011
7. NREL (2009) Opportunities and challenges for development of a mature concentrating photovoltaic power industry. http://nrel.gov/pv/pdfs/43208.pdf. Accessed 29 July 2011
8. MacKay D (2008) Sustainable energy—without the hot air. UIT Cambridge
9. Fthenakis V, Alsema E (2005) Photovoltaics energy payback times, greenhouse gas emissions and external costs 2004–early 2005 status. Wiley InterScience. doi:10.1002/pip.706
10. Mills A (2009) Spatial and temporal scales of solar variability: implications for grid integration of utility-scale photovoltaic plants. http://uwig.org/pvwork/11-Mills-Geographic DiversityAndPV.pdf. Accessed 29 July 2011
11. NREL (2008) Solar energy technology program, concentrating solar power. http://nrel.gov/csp/pdfs/43685.pdf. Accessed 29 July 2011
12. Mills D, Morgan R (2007) Solar thermal power as the plausible basis of grid supply. http://ausra.com/pdfs/T_1_1_David_Mills_2049.pdf. Accessed 29 July 2011
13. NEEDS (2008) Final report on technical data, costs, and life cycle inventories of solar thermal power plants. http://needs-project.org/docs/results/RS1a/RS1a%20D12.2%20Final%20report%20concentrating%20solar%20thermal%20power%20plants.pdf. Accessed 21 July 2011
14. Odeh S, Behnia M, Morrison G (2003) Performance evaluation of solar thermal electric generation systems. Energy Conversion and Management Sept 2003
15. Intergovernmental Panel on Climate Change (2005) Carbon dioxide capture and storage. http://ipcc.ch/publications_and_data/publications_and_data_reports.shtml. Accessed July 31 2011
16. Australian Resources and Investment (March 2010) http://info.australianresourcesand investment.com.au/default.aspx. Accessed 31 July 2011
17. Centre for Low Carbon Futures (2011) Carbon capture and utilisation in the green economy, using CO_2 to manufacture fuel, chemicals and materials. http://lowcarbonfutures.org/assets/media/clcf_ccu_report_21_july_final_sm.pdf.pdf. Accessed 23 July 2011
18. ITER (2010) International thermonuclear experimental reactor. http://iter.org/default.aspx. Accessed 2 Aug 2011
19. EWEA (2010) Wind energy and electricity prices. http://www.ewea.org/fileadmin/ewea_documents/documents/publications/reports/MeritOrder.pdf. Accessed 4 Aug 2011

Index

M. Nicholson, *The Power Makers' Challenge*, Green Energy and Technology,
DOI: 10.1007/978-1-4471-2813-7, © Springer-Verlag London 2012

MIX
Papier aus verantwortungsvollen Quellen
Paper from responsible sources
FSC® C116 358

If you have any concerns about our products,
you can contact us on
ProductSafety@springernature.com

In case Publisher is established outside the EU,
the EU authorized representative is:
Springer Nature Customer Service Center GmbH
Europaplatz 3, 69115 Heidelberg, Germany

Printed by Libri Plureos GmbH
in Hamburg, Germany